A Short Course in Fluid Mechanics

Thomas J. R. Hughes
and
Jerrold E. Marsden

Division of Structural Engineering and Structural Mechanics
and Department of Mathematics
University of California, Berkeley

ISBN 0-914098-15-2
Library of Congress Catalog Card Number: 76-6802

PUBLISH OR PERISH, INC.
6 BEACON STREET
BOSTON, MASS. 02108 (U.S.A.)

In Japan distributed exclusively by
KINOKUNIYA BOOK-STORE CO., LTD.
TOKYO, JAPAN

A SHORT COURSE IN FLUID MECHANICS

"All of the true things I am
about to tell you are shame-
less lies."

Books of Bokonon (Kurt Vonnegut)

Preface

These notes, from a one-quarter course given at Berkeley in 1974, provide a rapid introduction to the basic theorems of fluid mechanics. The presentation is rigorous, although we state but do not include proofs of any of the difficult theorems (such as existence theorems). On the other hand we do not hesitate to discuss interesting physical situations, or engage in speculations concerning interesting and little understood flows.

The presentation is not exhaustive. We present a number of results which we feel are fundamental, so the beginner can get into the subject quickly. Furthermore, we hope the notes may also be useful to engineers who feel a little uneasy about the foundations of their subject, or who wish to see some real difficulties which they may have overlooked. A few of the topics are presented in a new way, not generally available in textbooks.

The material in Sections 1-17 is basic and should be mastered by everyone wishing to understand fluid mechanics. The material gets more advanced from 18 through 23 and indeed the last two sections become speculative and chatty due to the primitive state of the theory.

The only background needed is a good course in vector calculus and, for Section 11, complex variables.

There are many standard books one can read for further discussion of points raised in these notes, or for supplementary material. A small sample of these is given below.

First of all, for additional examples, "paradoxes" and foundational points, see

[1] G. Birkhoff, "Hydrodynamics, a Study in Logic, Fact and Similtude," Princeton University Press (1960).

Some good general references are

[2] S.L. Goldstein, "Modern Developments in Fluid Mechanics," (2 volumes) Dover (1965),

[3] G.K. Batchelor, "An Introduction to Fluid Mechanics," Cambridge University Press (1970).

[4] K.O. Friedrichs and R. Von Mises, "Fluid Dynamics," Springer Lecture Notes in Applied Mathematical Sciences, Vol. 5, (1971),

[5] L.D. Landau and E.M. Lifschitz, "Fluid Mechanics," Addison Wesley (1959).

[6] R.E. Meyer, "Introduction to Mathematical Fluid Dynamics," Wiley (1971).

[7] H. Lamb, "Hydrodynamics," Dover (1945).

[8] L.M. Milne-Thompson, "Theoretical Hydrodynamics," Macmillan (1957).

[9] R. Courant and K. Friedrichs, "Supersonic Flow and Shock Waves," Interscience (1948).

[10] H. Schlichting, "Boundary Layer Theory," McGraw Hill (1960).

[11] J. Serrin, Mathematical Principles of Classical Fluid Mechanics in "Handbuch der Physik," Vol. VIII/1, Springer-Verlag (1959).

[12] O.A. Ladyzhenskaya, "The Mathematical Theory of Viscous Incompressible Flow," Gordon and Breach (1969).

[13] M. Shinbrot, "Lectures on Fluid Mechanics," Gordon and Breach (1973).

[14] A. Chorin, "Lectures on Turbulence Theory," Publish or Perish (1975).

[15] R. Aris, "Vectors, Tensors and the Basic Equations of Fluid Mechanics," Prentice-Hall (1962).

[16] J.A. Owczarek, "Introduction to Fluid Mechanics," International Textbook Co. (1968).

References [2-8], [15] are medium level basic texts. [9] and [10] are more specialized. [11-14] are quite mathematical, and [16] is a sample engineering oriented text. At this level, we found [2] and [11] particularly useful. Some further references on specific points are given in the text.

We thank M. McCracken and A. Chorin for help with §§21, 22, 23, P. Renz for pointing out some interesting references, and Harold Raffill for his fine job of typing the manuscript. Lee Rambeau prepared the artwork.

Contents

§1. Terminology; Velocity Fields, Particle Paths, Spatial and Material Coordinates.

In order to talk about some examples of fluid motions (water, air) we shall introduce some terminology.

Let Ω be a region in $\mathbb{R}^3$ (or sometimes we work in $\mathbb{R}^2$). By "region" we shall always mean a "nice region" with smooth (or occasionally piecewise smooth where appropriate) boundary $\partial\Omega$. For instance, regions Ω which are bounded by graphs of smooth functions will be adequate for our purposes. For these there is no doubt about what is meant by the boundary, the inside, or the outward unit normal, etc.

Imagine a fluid moving in Ω. See Figure 1-1. Each particle of fluid follows a certain trajectory. Thus for each $\underline{x} \in \Omega$, there is a path $\underline{\sigma}(t) \in \Omega$ representing the trajectory of $\underline{x}$.

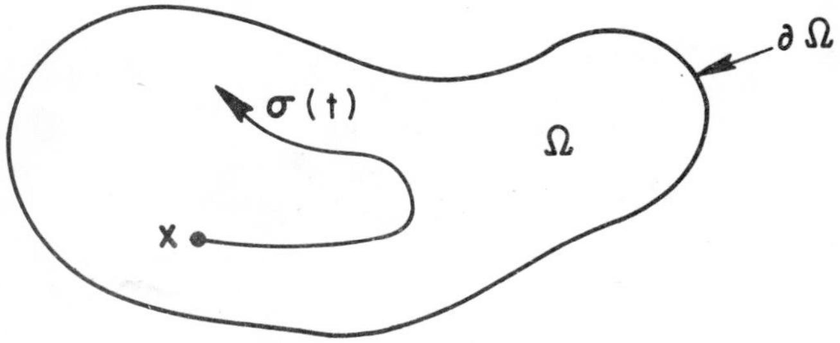

FIG. 1-1

In order to label the different trajectories, let us write $\underline{\phi}(t, \underline{x})$ for the path followed by $\underline{x}$, with the initial condition $\underline{\phi}(0, \underline{x}) = \underline{x}$.

For fixed t, we write $\underline{\phi}_t(\underline{x}) = \underline{\phi}(t, \underline{x})$, so the mapping

$\phi_t : \Omega \to \Omega$ sends x to the point to which x has flowed after time t. Thus ϕ_0 is the identity.

Let v denote the velocity of the flow (see Figure 1-2). Thus at $y = \phi(t, x)$, v is the velocity vector of the curve $\sigma(t) = \phi_t(x)$; i.e. $v = \sigma'(t)$. In other words:

$$v(t, \phi(t, x)) = \frac{d}{dt}\phi(t, x) \tag{1}$$

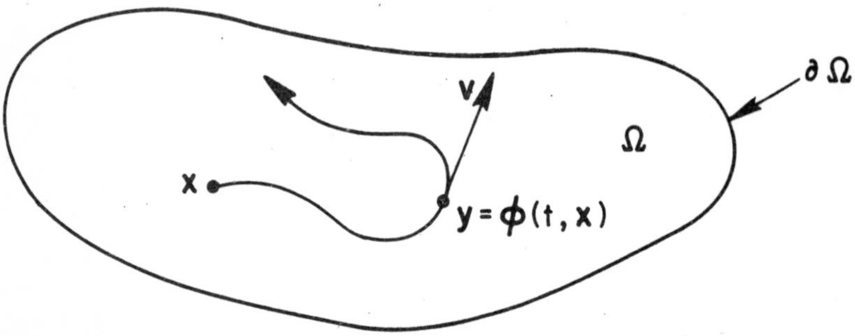

FIG. 1-2

Thus we get a time dependent velocity field $v(t, y)$ on Ω, provided $\phi_t : \Omega \to \Omega$ is one-to-one and onto.

The paths $\sigma(t) = \phi(t, x)$ are called <u>particle paths</u>. These are directly observable.

<u>Streamlines</u> $s(t)$ are obtained by freezing t at t_0 (as in a snapshot) and solving the differential equations

$$s'(t) = v(t_0, s(t)) . \tag{2}$$

Streamlines coincide with particle paths if the flow is stationary; i.e. $\underline{v}$ does not depend explicitly on t. This is obvious by comparison of (1) and (2).

The flow is incompressible if ϕ_t preserves volumes. In §5 we shall show this is equivalent to $\operatorname{div} \underline{v} = 0$ (at each t). Most low or medium velocity flows of air and water (including subsonic airplane flight) may be viewed as incompressible for practical calculations.

The spatial (or Eulerian*) point of view means that we regard functions as being of t and $\underline{x} \in \Omega$ and have the fluid rush by. (This is intentionally vague.) The material (or Lagrangian) point of view means that we regard functions as being functions of t and of the fluids initial position.

An example will clarify this: if $f(t, \underline{x})$ is a function on Ω, then we regard f as being in spatial coordinates. The corresponding function in material coordinates is defined by

$$F(t, \underline{x}) = f(t, \underline{\phi}(t, \underline{x})) . \tag{3}$$

Above we defined the velocity field $\underline{v}(t, \underline{x})$. It is in spatial coordinates. The corresponding field in material coordinates is

$$\underline{V}(t, \underline{x}) = \underline{v}(t, \underline{\phi}(t, \underline{x})) . \tag{4}$$

---

*The historical credits are often debated. See C. Truesdell, "Essays on the History of Mechanics," Springer (1968) for more information.

From (1) we see the geometric significance of (4); see Figure 1-3.

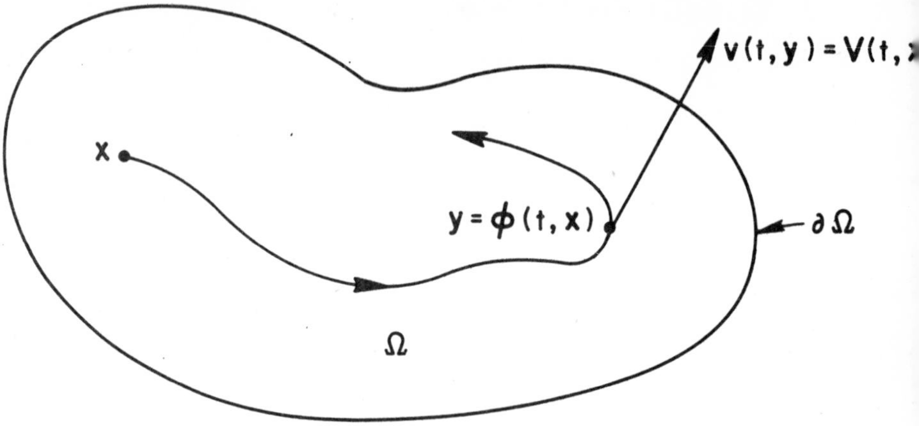

FIG. 1-3

Problem. Compute the particle paths for

(a) $\underline{v}(x, y) = -y\underline{i} + x\underline{j}$

and (b) $\underline{v}(x, y) = \frac{-y}{x^2+y^2}\,\underline{i} + \frac{x}{x^2+y^2}\,\underline{j}$.

§2. Some Examples of Fluid Motions.

We briefly discuss a few simple experiments which can be done in a kitchen or by observation of the weather.

1. Stirring Tea. Place tea leaves in a clear glass of water and stir. As everyone knows, the tea leaves tend to gather at the center, on the bottom of the glass. Why is this?

The "true explanation" is in reality complicated if one wants to use the equations of fluid mechanics (developed later). However, the general features can be understood here by qualitative arguments. We caution the reader that such arguments often are erroneous and lead to false predictions (see examples in §10 and in the book of Birkhoff [1] mentioned in the Preface).

The reasoning often given goes as follows: the rotary action tends to push the fluid to the sides of the container by centrifugal force. This is less immediately along the walls of the glass because the fluid

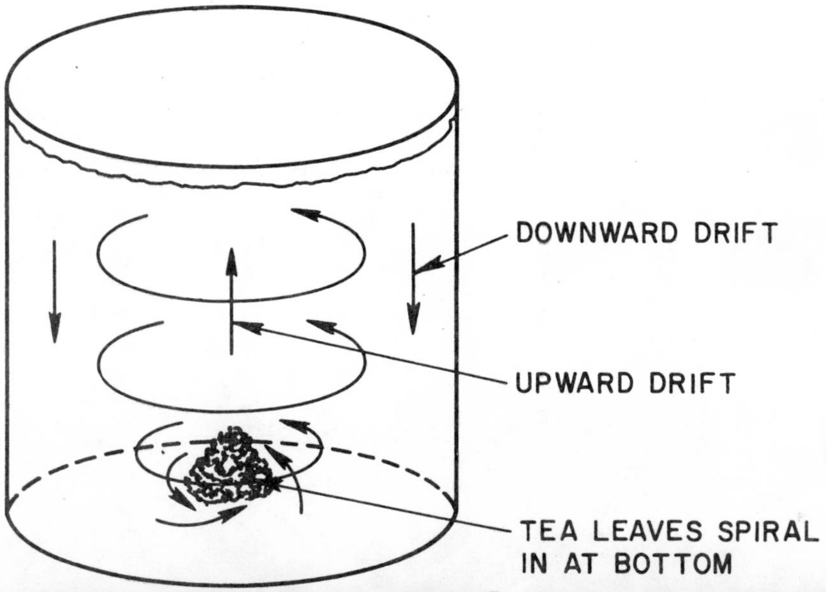

FIG. 2-1

adheres to the glass. Since water is incompressible, it can't all be pushed to the sides, so it comes in along the very bottom of the glass, taking the tea leaves with it.* See Figure 2-1.

However, our intuition and this type of explanation break down in the more severe and controlled situation of <u>Couette flow</u>. Here fluid is placed between two rotating cylinders (see Figure 2-2). When these cylinders are rotated, again circulations are set up like the teacup model, if the rotations are large enough. But it does so in a discrete number of bands, called <u>Taylor cells</u>. (For details on the experiment

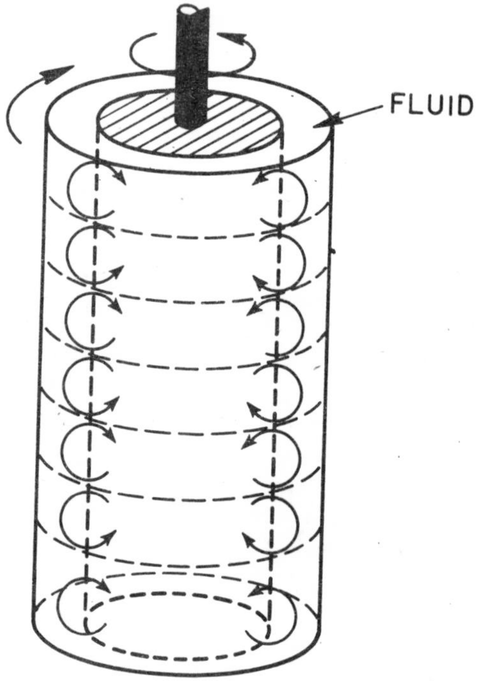

FIG. 2-2

---

*Surface tension also has an effect. If the fluid is rotating quickly, the concavity of the rotating surface has an effect as well. An explanation by A. Einstein given in 1926 is similar to the one given here (see A. Einstein, "Ideas and Opinions," edited by C. Seelig, Crown (1954) p. 249).

and to see how complex it really is, see D. Coles, Transition in Circular Couette Flow, Journal of Fluid Mechanics, 21 (1965) 385-425.)

2. Pouring Experiments. An easy way of setting up interesting flows is to pour or drop one liquid carefully into another, such as cream in tea or ink in water. Typically, small circulation whirlpools, or rotations of fluid are set up in interesting patterns (see Figure 2-3).

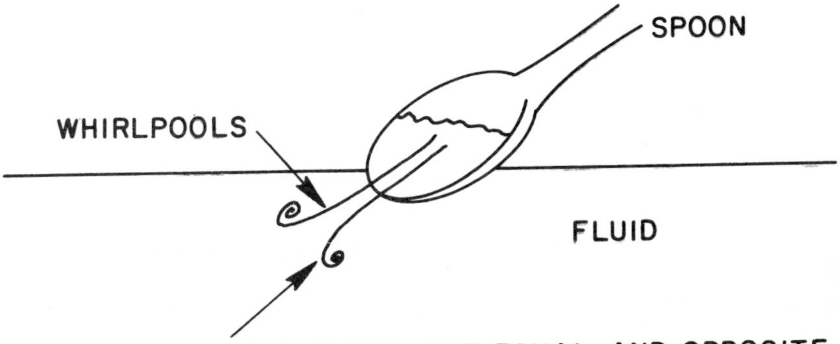
SPOON
WHIRLPOOLS
FLUID
THE TWO CIRCULATIONS ARE EQUAL AND OPPOSITE,
SO THE NET CIRCULATION = 0.

FIG. 2-3

In each case, one observes a basic principle* of fluid mechanics (to be established in §14), namely conservation of circulation. Before pouring, the net circulation is zero, and it is after as well, although it may be non-zero locally.

A similar phenomenon occurs in aircraft flight. Behind large jets, substantial vortices are formed; always in equal and opposite pairs (Figure 2-4) (this is not due to the engines).

---

*The principle referred to, Kelvin's circulation theorem, holds under specific conditions spelled out in §§14, 16.

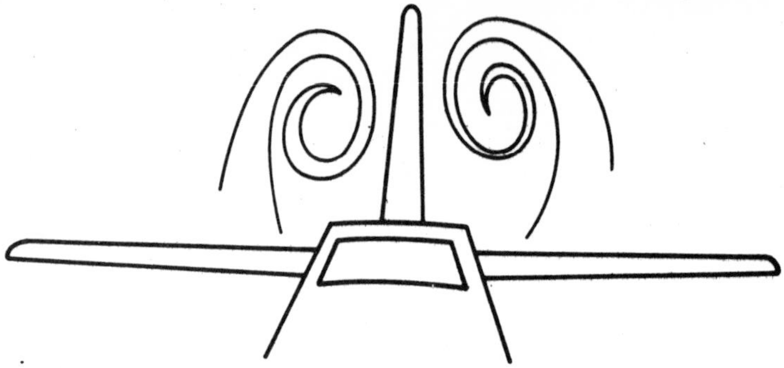

VIEW LOOKING TOWARDS
REAR OF AIRPLANE

FIG. 2-4

Conservation of circulation is of obvious importance in the weather, especially in the formation of tornados. These often occur near the edge of a rapidly advancing cold front (Figure 2-5).

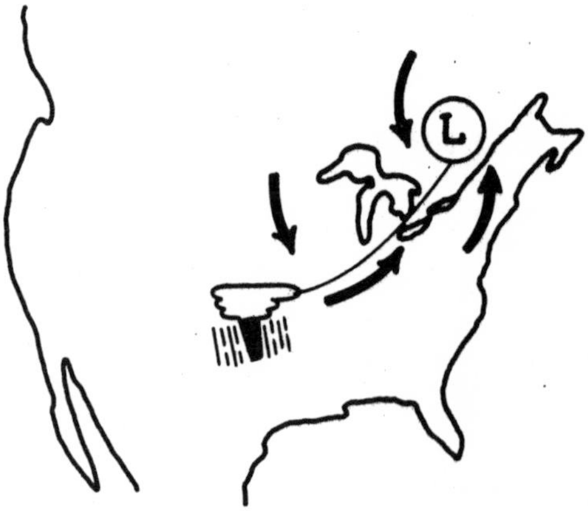

FIG. 2-5

Because of the wind patterns, a small amount of counterclockwise rotation exists near the front. During a thunderstorm, rapidly rising air gathers this circulation to a small region. The circulation becomes violent due to its concentration and conservation of circulation. Tornados can last a good while, especially if they are vertically stretched, so as to become smaller in cross-section (incompressibility!). Of course, this can be maintained only so long due to frictional losses.

Conservation of circulation is also seen in the flow behind an obstacle. The flow there assumes, at the right ranges of velocities, the form in Figure 2-6; the famous "Karmen vortex street."

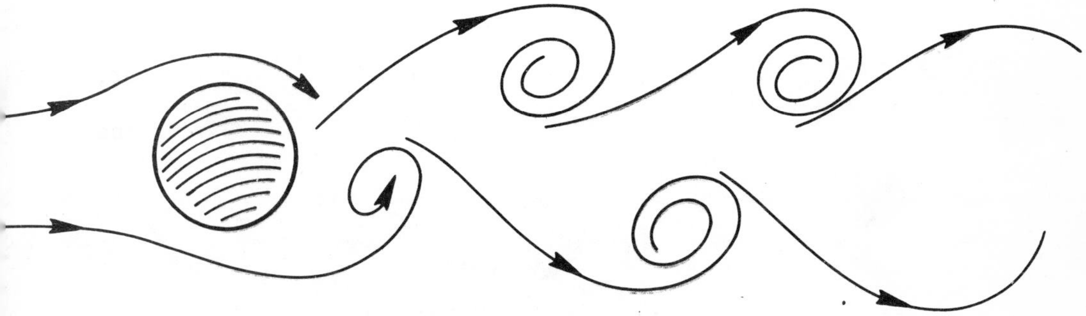

FIG. 2-6

Such patterns are very difficult to explain theoretically and only recently have been obtained numerically (see A. Chorin, Numerical Study of Slightly Viscous Flows, Journal of Fluid Mechanics 57 (1973) 785-796). It is patterns like this that enable whistles to work! (See Scientific American, January 1970.)

Research Project. Clouds in the sky often form into long rolls or cells.

Simulate this by heating, in a pan, a layer of chocolate overlaid by milk. You should observe rolls. Hexagons form if oil on the surface of water is used (surface tension is needed). This is referred to as the "Bénard phenomenon" and is discussed in several books.†

Problem. In the wintertime, after a fresh snow and a north wind, one observes a hollow in the snow on the north side of trees. Using Figure 2-7, conservation of net circulation and incompressibility, try to explain this.

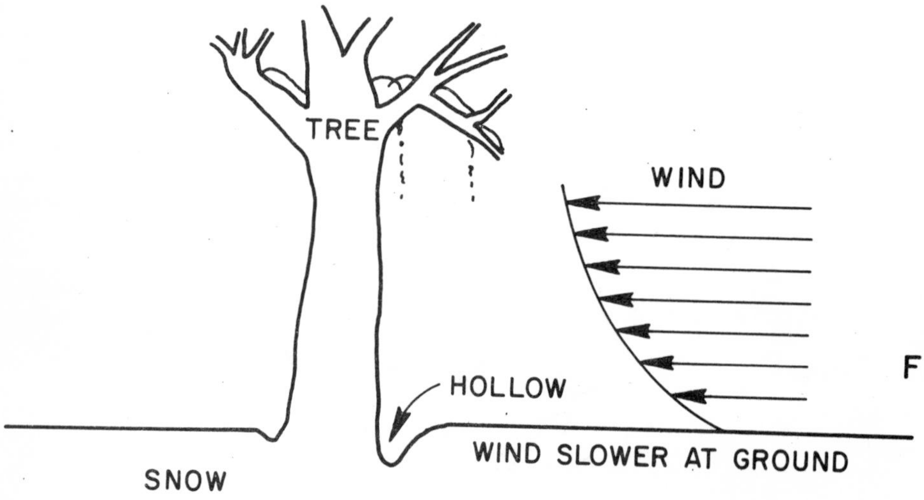

FIG. 2-

We hope these examples stimulate the reader into believing that some genuinely exciting flows and many open problems exist in fluid mechanics. The subject is far from closed, despite its age and, in a

---

†For a nice account, see Cellular Convection by J.A. Whitehead, American Scientist 59 (1971) 444. See also Observations of an Early Morning Cup of Coffee, by V.J. Schaefer, American Scientist 59 (1971) p. 534. The fact that surface tension is needed to get hexagons is due to A. Chorin (A Numerical Method for Solving Incompressible Viscous Flow Problems, Journal Comp. Phys. 2 (1967) 12-26). He used the equations neglecting surface tension and showed that rolls arise but hexagons do not.

ense, it is just beginning. For instance, we are only barely beginning o understand the real mechanisms involved in boundary layer theory the motion of fluids near walls) and in turbulence theory (the chaotic otion of fluids), as well as the deeper significance of the general heory.*

---

*See, e.g., A. Chorin [14], D. Ebin and J. Marsden, Groups of Diffeomorphisms and the Motion of an Incompressible Fluid, Ann. of Math. 92 (1970) 102-163, and J. Marsden, "Applications of Global Analysis in Mathematical Physics," Publish or Perish, (Lecture Notes #2) (1974). Some of these points will be elaborated on in Sections 21-23.

§3. Some Preliminary Mathematics; An Algebraic Lemma and a Review of Vector Calculus.

The following will be useful to us later. We record it here as a simple result in matrix algebra.

An Algebraic Lemma. *Let* $\underline{A}$ *be an* $n \times n$ *matrix such that for any unit vector* $\underline{n}$, $\underline{A}\cdot\underline{n}$ *is parallel to* $\underline{n}$. *Then* $\underline{A} = p\,\underline{I}$ *where* p *is a real number.*

Proof. Let $\{\underline{e}_i\}_1^n$ be an orthonormal basis for $\mathbb{R}^n$. Then, by hypothesis, there is a number p_i such that $\underline{A}\underline{e}_i = p_i\underline{e}_i$ for each $i = 1, \ldots, n$. The components a_{ij} of $\underline{A}$ with respect to the basis $\{\underline{e}_i\}_1^n$ are given by

$$\begin{aligned} a_{ij} &= \underline{e}_i\cdot(\underline{A}\underline{e}_j) \\ &= \underline{e}_i\cdot(p_j\underline{e}_j) \\ &= p_j\underline{e}_i\cdot\underline{e}_j \\ &= p_j\delta_{ij} \end{aligned}$$

where δ_{ij} is the Kronecker delta, i.e. $\delta_{ij} = 1$ if $i = j$, $\delta_{ij} = 0$ if $i \neq j$.

Now $p_i = \underline{e}_i\cdot\underline{A}\underline{e}_i$, so if we let $\underline{A}(\underline{e}_i + \underline{e}_j) = p_{i+j}(\underline{e}_i + \underline{e}_j)$, $i \neq j$ and take the inner product with $\underline{e}_i$,

$$\begin{aligned} p_{i+j} &= \underline{e}_i\cdot\underline{A}(\underline{e}_i + \underline{e}_j) \\ &= \underline{e}_i\cdot(p_i\underline{e}_i + p_j\underline{e}_j) \\ &= p_i \;. \end{aligned}$$

Similarly $p_{i+j} = p_j$. Thus $p_i = p_j$ for all i, j , so $a_{ij} = p\delta_{ij}$. ■

Review of Vector Calculus.*

Let $\underline{v} = (v_1, v_2, v_3)$ be a vector field defined on a region $\Omega \subset \mathbb{R}^3$. The divergence of $\underline{v}$ is given by

$$\text{div } \underline{v} = \underline{\nabla} \cdot \underline{v} = \sum_{i=1}^{3} \frac{\partial v_i}{\partial x_i}$$

and the curl of $\underline{v}$ is given by

$$\text{curl } \underline{v} = \underline{\nabla} \times \underline{v} = \begin{vmatrix} \underline{i} & \underline{j} & \underline{k} \\ \frac{\partial}{\partial x} & \frac{\partial}{\partial y} & \frac{\partial}{\partial z} \\ v_1 & v_2 & v_3 \end{vmatrix}$$

Our vector fields will always be at least C^1 , so these operations make sense.

A curve $\underline{\sigma} : [a, b] \to \mathbb{R}^3$ is a C^1 (or piecewise C^1) mapping. The image of $\underline{\sigma}$ is denoted by C . If $\underline{v}$ is a vector field defined on $\underline{\sigma}$, then we write

$$\int_C \underline{v} = \int_C \underline{v} \cdot \underline{ds} = \int_a^b \underline{v}(\underline{\sigma}(t)) \cdot \underline{\sigma}'(t)dt ,$$

where $\underline{\sigma}'(t)$ denotes the velocity of $\underline{\sigma}$ (see Figure 3-1). By the change of variables formula, the number $\int_C \underline{v}$ is unaffected by a

---

*Again Ω is a "nice region". See, e.g., J. Marsden and A.Tromba "Vector Calculus", W.H. Freeman and Co. (1976) for details of the summary given here.

reparametrization of $\underline{\sigma}$. When C is a loop (Figure 3-2), $\int_C \underline{v}$ is called the <u>circulation</u> of $\underline{v}$ around C .

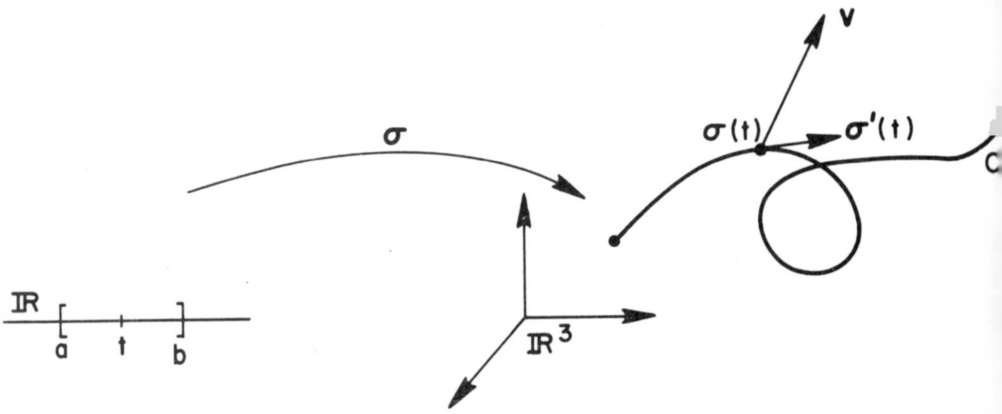

FIG. 3-1

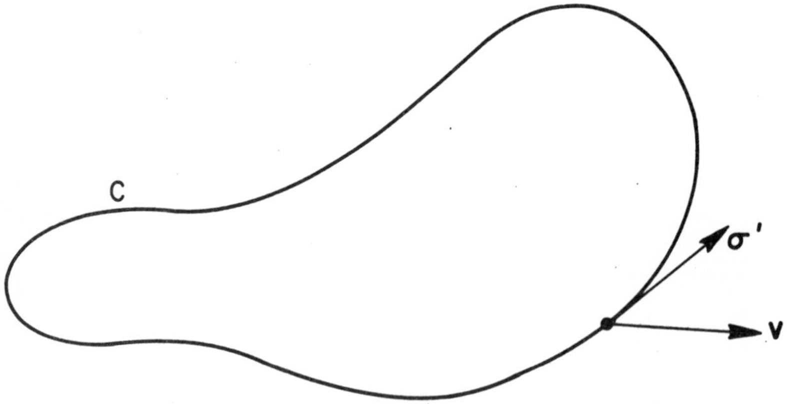

FIG. 3-2

Let $D \subset \mathbb{R}^2$ be a region in the plane (e.g., a rectangle). A <u>parametrized surface</u> $\underline{\Phi}: D \to \mathbb{R}^3$ is a C^1 (or piecewise C^1) mapping.

The image of $\underline{\Phi}$ is denoted by S . If $\underline{v}$ is a vector field defined over S we write

$$\int_S \underline{v} = \int_S \underline{v}\cdot\underline{dS} = \iint_D \underline{v}(\underline{\Phi}(u, v))\cdot(\underline{T}_u \times \underline{T}_v)\,du\,dv ,$$

where the parameters u, v are the cartesian coordinates on $\mathbb{R}^2$ and

$$\underline{T}_u = \frac{\partial\Phi}{\partial u} , \quad \underline{T}_v = \frac{\partial\Phi}{\partial v}$$

are vectors tangent to the images of the u, v coordinate lines on S , respectively (see Figure 3-3). As above, the number $\int_S \underline{v}$ is independ-

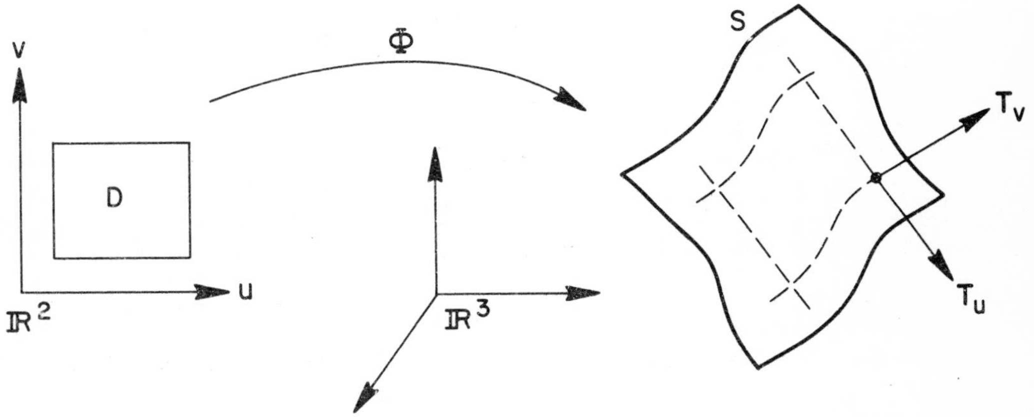

FIG. 3-3

ent of the parametrization of S . If $\underline{v}$ is the velocity vector field of a fluid, then $\int_S \underline{v}$ is the volume flow per unit time through S . In any case $\int_S \underline{v}$ is called the <u>flux of</u> $\underline{v}$ across S . Recall that $\underline{dS} = \underline{T}_u \times \underline{T}_v\ du\,dv$ is the differential surface area element for S . We also write $da = \|\underline{T}_u \times \underline{T}_v\|dudv$ for the area element as a scalar.

We can similarly integrate over surfaces S which are made up of a union of a number of parametrized surfaces, by addition over these pieces.

Stokes Theorem. *If* S *is a surface with boundary* C , *oriented* as illustrated in Figure 3-4 and* $\underline{v}$ *is a vector field on* S , *then*

$$\int_S (\underline{\nabla} \times \underline{v}) \cdot \underline{dS} = \int_C \underline{v} \cdot \underline{ds} \ .$$

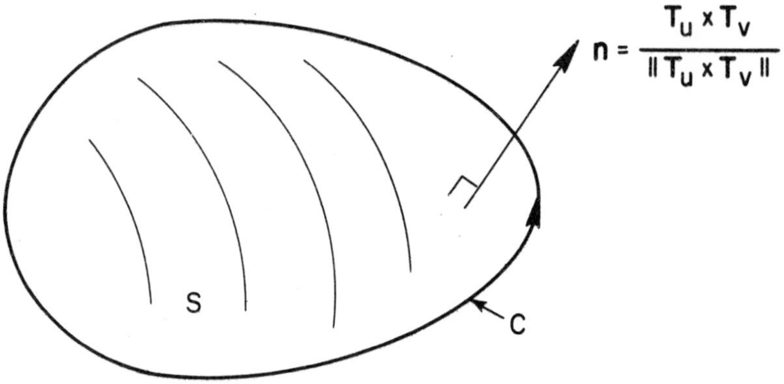

FIG. 3-4

From the mean value theorem for integrals, we get:

Corollary. *Let* $\underline{n} = \underline{T}_u \times \underline{T}_v / \|\underline{T}_u \times \underline{T}_v\|$ *and suppose* S *is a disc* D_r *of radius* r *with boundary* C_r . *Let* p *denote the center of* D_r *and* $A(D_r)$ *its area* (Figure 3-5). *Then*

$$(\underline{\nabla} \times \underline{v}) \cdot \underline{n} \ (p) = \lim_{r \to 0} \frac{1}{A(D_r)} \int_{C_r} \underline{v} \ .$$

---

*Again, we assume $\underline{v}$ is C^1 and the regions are always "nice", as in §1, so there is no doubt as to what the orientations are.

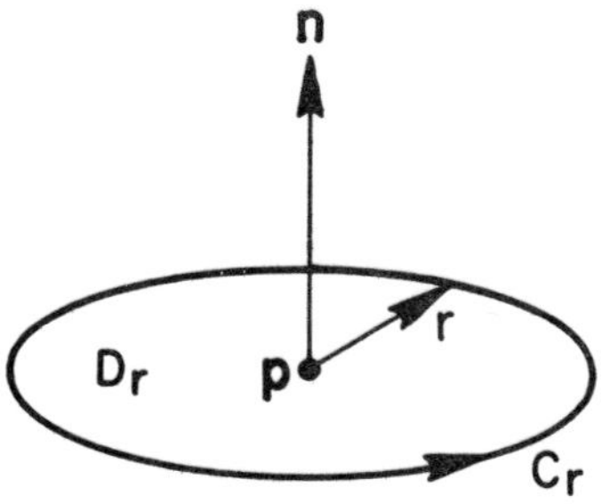

FIG. 3-5

It is in this sense that one says that "the curl of a vector field is its circulation per unit area."

Gauss Theorem. *Let* W *denote a volume contained in* $\mathbb{R}^3$ *with boundary surface* S *and let* $\underline{n}$ *denote the unit outward normal vector field to* S . *Then* $\int_W \underline{\nabla}\cdot\underline{v}\, dx = \int_S \underline{v}\cdot\underline{dS}$.

Again the mean value theorem yields:

Corollary. *Let* W *be the ball* B_r *of radius* r *with boundary sphere* S_r *and let* p *be the center of* B_r *and* $V(B_r)$ *be the volume of* B_r . *Then*

$$(\underline{\nabla}\cdot\underline{v})(p) = \lim_{r\to 0} \frac{1}{V(B_r)} \int_{S_r} \underline{v} \,.$$

One thus says that "the divergence of a vector field is its flux per unit volume."

Physical Interpretations. Let $\underline{v}$ be the velocity vector field of a fluid.

Orient an "infinitesimal paddle wheel" in the direction $\underline{n}$, say pointing outward from the page. Then if $(\underline{\nabla} \times \underline{v})\cdot\underline{n}$ is $+$, $-$, the paddle wheel will spin in the counter-clockwise, clockwise direction, respectively. If $(\underline{\nabla} \times \underline{v})\cdot\underline{n} = 0$, the paddle wheel will remain stationary (see Figure 3-6).

If $\underline{\nabla}\cdot\underline{v}$ is $+$, $-$, at a point, then the fluid is locally expanding, contracting, respectively. If $\underline{\nabla}\cdot\underline{v} = 0$ the motion is volume-preserving.

<u>Problem</u>. Let $\vec{\nabla}\cdot\vec{v} = 3y$

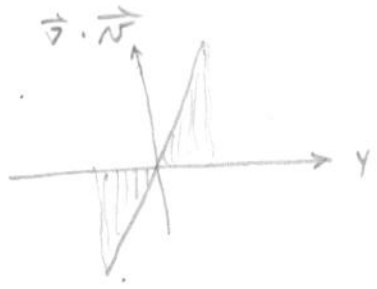

to every contribution there is an equal and opposite contribution

$$\underline{v}(x, y, z) = 3xy\underline{i} + xz^2\underline{j} + y^3\underline{k} \ . \quad \Rightarrow \text{flux} = 0$$

Compute the flux of $\underline{v}$ through the sphere of ~~unit~~ any radius.

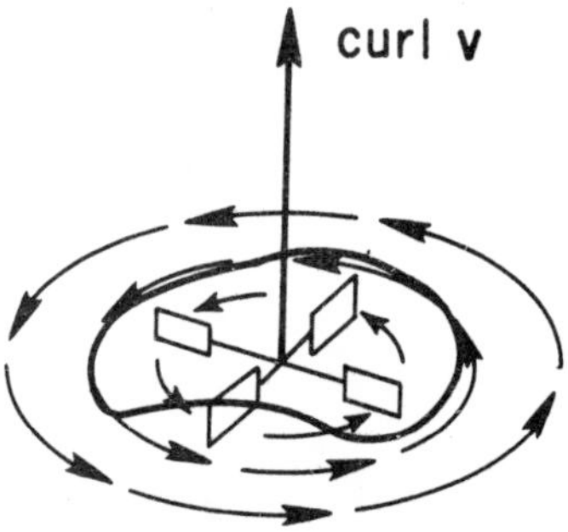

FIG. 3-6

We also recall that the <u>gradient</u> of a function f is defined by

$$\text{grad } f = \underline{\nabla} f = \frac{\partial f}{\partial x}\underline{i} + \frac{\partial f}{\partial y}\underline{j} + \frac{\partial f}{\partial z}\underline{k} \ .$$

For f of class C^2 we have $\underline{\nabla} \times \underline{\nabla} f = 0$ and $\underline{\nabla} \cdot (\underline{\nabla} \times \underline{v}) = 0$ for $\underline{v}$ of class C^2.

Problem. Let $f(x, y, z) = x^2 y e^z$ and let C be the curve $\underline{\sigma}(t) = (\sin t, \cos t, t)$, $0 \leqslant t \leqslant 2\pi$. Evaluate $\int_C \underline{\nabla} f$.

(Hint: Use the chain rule and the Fundamental Theorem of Calculus).

Problem. Review the proof that if $\underline{v}$ is defined on all of $\mathbb{R}^3$ and curl $\underline{v} = 0$ then $\underline{v} = \underline{\nabla} f$ for some f. Is it true if $\underline{v}$ is not defined everywhere?

A summary of important vector identities as well as the formulae for $\underline{\nabla} \times \underline{v}$, $\underline{\nabla} \cdot \underline{v}$, etc. in cylindrical and spherical coordinates are given in the appendix. They will be used in succeeding sections.

Problem. Let E be a vector space (not necessarily finite dimensional) and $T: E \to E$ a linear transformation. Assume $Tx = p(x)x$ for some scalar function on E. Prove $p(x)$ is constant.

§4. Flow of a Vector Field.

Let $\underline{v}(t, \underline{x})$ be a time dependent vector field defined for $\underline{x} \in \Omega$ and $t \in \mathbb{R}$. An integral curve for $\underline{v}$ is a (differentiable) curve $\underline{\sigma}(t)$, such that

$$\underline{\sigma}'(t) = \underline{v}(t, \underline{\sigma}(t)) .$$

We may think of $\underline{v}$ as the velocity vector field of a fluid and $\underline{\sigma}$ as a particle path as explained in §1.

Basic Existence Theorem. *Assume* $\underline{v}(t, \underline{x})$ *is a* C^1 *function of* $\underline{x}$ *and* t. *Then, given* $\underline{x}_0$ *and* t_0, *there exists a unique integral curve* $\underline{\sigma}(t)$ *defined on some small* t-*interval about* t_0, *such that* $\underline{\sigma}(t_0) = \underline{x}_0$.

See any book on ordinary differential equations, such as P. Hartman, "Ordinary Differential Equations," Wiley (1964).

Example. Assume $\Omega = \mathbb{R}$, $v(t, x) = x^2$, $t_0 = 0$ and $x_0 = 1$. Then the solution of $\sigma' = \sigma^2$ is $\sigma = 1/(1 - t)$. Note that the solution cannot be extended beyond $t = 1$ since $\sigma \uparrow +\infty$ as $t \uparrow 1$.

On the other hand, the following holds:

Theorem. *If* Ω *has a smooth boundary* $\partial\Omega$ *and* $\underline{v}$ *is parallel to* $\partial\Omega$, *then integral curves for* $\underline{v}$ *starting in* Ω *remain in*

Ω. *If Ω is bounded, integral curves are defined for as long as $\underline{v}$ is defined.*

This is proved by standard methods in differential equations (see the above reference).

Flow of $\underline{v}$. Assume $\underline{v}$ is parallel to $\partial\Omega$. By the above theorem, integral curves for $\underline{v}$ starting in Ω, remain in Ω. The flow of $\underline{v}$ is a mapping $\underline{\phi}: \Omega \to \Omega$ such that $\underline{\phi}_t(\underline{x}_0) = \underline{\sigma}(t)$, where $\underline{\sigma}$ is the integral curve for $\underline{v}$ passing through $\underline{x}_0$ at $t = 0$. Clearly $\underline{\phi}_0 = \text{id}$ (the identity map on Ω), and, as in §1, $\frac{\partial}{\partial t}\underline{\phi}(t, \underline{x}) = \underline{v}(t, \underline{\phi}(t, \underline{x}))$.

Also, from the theory of ordinary differential equations, one can show that if $\underline{v}$ is a C^k vector field, $k \geqslant 1$, then $\phi(t, \underline{x})$ is also C^k; so we can freely differentiate it. (It is already clear that ϕ is C^{k+1} in t, but is not so obvious in x; it expresses the smooth dependence of the solution of a differential equation on the initial conditions.)

Problem. Suppose $\underline{v}$ is independent of t. Then using uniqueness of solutions, prove that $\underline{\phi}_t$ satisfies $\underline{\phi}_{t+s} = \underline{\phi}_t \circ \underline{\phi}_s$ (i.e. $\underline{\phi}(t + s, \underline{x}) = \underline{\phi}(t, \underline{\phi}(s, \underline{x}))$).

Problem. Suppose $f(t, \underline{x})$ is a scalar function and satisfies $\frac{\partial f}{\partial t} + \underline{\nabla f} \cdot \underline{v} = 0$ ($\underline{\nabla f}$ is computed holding t fixed). Prove that

$$f(t, \phi(t, \underline{x})) = f(0, \underline{x})$$

"f travels with the flow"

for all t.

§5. The Transport Theorem.

Now we prove a very useful identity involving functions and flows. It will be completely basic in later sections.

The Transport Theorem. *Let* $\underline{v}(t, \underline{x})$ *be a* C^2 *vector field on* Ω *, parallel to* $\partial\Omega$ *, with flow* $\underline{\phi}(t, \underline{x})$ *, and let* $f(t, \underline{x})$ *be a function on* Ω *. Assume that* $\underline{\phi}$ *is invertible as a function of* $\underline{x}$ *for a range of* t *. Then in this range*

$$\frac{d}{dt}\int_{\underline{\phi}_t(W)} f(t, \underline{x})dx = \int_{\underline{\phi}_t(W)} \left(\frac{\partial f}{\partial t} + \underline{v}\cdot\underline{\nabla} f + f\underline{\nabla}\cdot\underline{v}\right)dx ,$$

where W *is any subregion of* Ω *with piecewise smooth boundary.*

Lemma. *Let* $J(t, \underline{x})$ *be the Jacobian determinant of* $\underline{\phi}(t, \underline{x})$ *, i.e.*

$$J(t, \underline{x}) = \begin{vmatrix} \frac{\partial\phi_1}{\partial x} & \frac{\partial\phi_2}{\partial x} & \frac{\partial\phi_3}{\partial x} \\ \frac{\partial\phi_1}{\partial y} & \frac{\partial\phi_2}{\partial y} & \frac{\partial\phi_3}{\partial y} \\ \frac{\partial\phi_1}{\partial z} & \frac{\partial\phi_2}{\partial z} & \frac{\partial\phi_3}{\partial z} \end{vmatrix}$$

Then if t *is in the range in question,* $J(t, \underline{x}) > 0$ *and*

$$\frac{\partial}{\partial t} J(t, \underline{x}) = J(t, \underline{x})(\operatorname{div} \underline{v})(t, \underline{x}) .$$

Proof. Since $\underline{\phi}(0, \underline{x}) = \text{id}$ on Ω and $\underline{\phi}$ is smooth, the continuity of the determinant map insures $J(t, \underline{x}) > 0$ for some small t-interval, and since it is not zero, it must remain > 0 on the given range.

The determinant J can be differentiated by recalling the fact that the determinant of a matrix is multilinear in the columns (or rows). Thus, holding $\underline{x}$ fixed throughout, we have

$$\frac{\partial}{\partial t} J(t, \underline{x}) = \begin{vmatrix} \frac{\partial}{\partial t}\frac{\partial \phi_1}{\partial x} & \frac{\partial \phi_2}{\partial x} & \frac{\partial \phi_3}{\partial x} \\ \frac{\partial}{\partial t}\frac{\partial \phi_1}{\partial y} & \frac{\partial \phi_2}{\partial y} & \frac{\partial \phi_3}{\partial y} \\ \frac{\partial}{\partial t}\frac{\partial \phi_1}{\partial z} & \frac{\partial \phi_2}{\partial z} & \frac{\partial \phi_3}{\partial z} \end{vmatrix}$$

$$+ \begin{vmatrix} \frac{\partial \phi_1}{\partial x} & \frac{\partial}{\partial t}\frac{\partial \phi_2}{\partial x} & \frac{\partial \phi_3}{\partial x} \\ \frac{\partial \phi_1}{\partial y} & \frac{\partial}{\partial t}\frac{\partial \phi_2}{\partial y} & \frac{\partial \phi_3}{\partial y} \\ \frac{\partial \phi_1}{\partial z} & \frac{\partial}{\partial t}\frac{\partial \phi_2}{\partial z} & \frac{\partial \phi_3}{\partial z} \end{vmatrix}$$

$$+ \begin{vmatrix} \frac{\partial \phi_1}{\partial x} & \frac{\partial \phi_2}{\partial x} & \frac{\partial}{\partial t}\frac{\partial \phi_3}{\partial x} \\ \frac{\partial \phi_1}{\partial y} & \frac{\partial \phi_2}{\partial y} & \frac{\partial}{\partial t}\frac{\partial \phi_3}{\partial y} \\ \frac{\partial \phi_1}{\partial z} & \frac{\partial \phi_2}{\partial z} & \frac{\partial}{\partial t}\frac{\partial \phi_3}{\partial z} \end{vmatrix}$$

Since $\underline{\phi}$ is the flow of $\underline{v}$, it is C^2. Thus we have

$$\begin{aligned} \frac{\partial}{\partial t}\frac{\partial \phi_i}{\partial x_k}(t, \underline{x}) &= \frac{\partial}{\partial x_k}\frac{\partial}{\partial t}\phi_i(t, \underline{x}) \\ &= \frac{\partial}{\partial x_k} v_i(t, \underline{\phi}(t, \underline{x})) \\ &= \sum_{j=1}^{3} \frac{\partial v_i}{\partial x_j}\frac{\partial \phi_j}{\partial x_k}, \end{aligned}$$

where we have used the chain rule and $i, k = 1, 2, 3$. Employing this result in the formula for $\partial J/\partial t$ yields the result. ■

Proof of the Transport Theorem. By the change of variables formula we can write

$$\int_{\underline{\phi}_t(W)} f(t, \underline{y})dy = \int_W f(t, \underline{\phi}(t, \underline{x}))J(t, \underline{x})dx .$$

Since the integrand of the right-hand side is C^1 and W does not change with t , we can differentiate under the integral sign. Then, using the chain rule and the lemma, we get

$$\frac{d}{dt}\int_W f(t, \underline{\phi}(t, \underline{x}))J(t, \underline{x})dx = \int_W \left(\frac{\partial}{\partial t} f + \underline{\nabla} f\cdot\underline{v} + f \operatorname{div} \underline{v}\right)J\, dx .$$

Changing variables back yields the result. ■

Material Derivative. A function f on Ω is said to be in spatial coordinates when it is written as $f(t, \underline{x})$, in material coordinates when it is written $f(t, \underline{\phi}(t, \underline{x}))$. The material derivative D/Dt of f is given by the formula

$$\frac{Df}{Dt} = \frac{\partial f}{\partial t} + \underline{\nabla} f\cdot\underline{v} .$$

Notice that $\frac{Df}{Dt}$ is exactly the t-derivative of $f(t, \underline{\phi}(t, \underline{x}))$.

Problem. Let $f, g: \mathbb{R} \times \Omega \to \mathbb{R}$ and $h: \mathbb{R} \to \mathbb{R}$ be C^1 mappings. Prove the following calculus formulas for D/Dt :

(i) $\frac{D}{Dt}(f + g) = \frac{Df}{Dt} + \frac{Dg}{Dt}$

(ii) $\frac{D}{Dt}(f \cdot g) = f\frac{Dg}{Dt} + g\frac{Df}{Dt}$ (Leibniz or product rule)

(iii) $\frac{D}{Dt}(h \circ g) = (h' \circ g)\frac{Dg}{Dt}$ (chain rule)

Definition. A flow $\underline{\phi}_t$ is called incompressible if

$$\frac{d}{dt}\int_{\underline{\phi}_t(W)} dx = 0$$

(i.e., if $\underline{\phi}_t$ preserves volumes).

Theorem. *A flow* $\underline{\phi}_t$ *is incompressible if and only if* $\operatorname{div} \underline{v} = 0$.

Proof. Set $f \equiv 1$ in the transport theorem. ∎

Problem. Use the transport theorem to establish the following formula of Reynolds:

$$\frac{d}{dt}\int_{\underline{\phi}_t(W)} f(t, \underline{x})dx = \int_{\underline{\phi}_t(W)} \frac{\partial f}{\partial t}(t, \underline{x})dx + \int_{\partial\underline{\phi}_t(W)} f\underline{v}\cdot\underline{dS} .$$

Interpret the result physically.

§6. Conservation of Mass.

We take it as a physical axiom that given a fixed volume of fluid W, throughout its motion $\phi_t(W)$, its mass remains constant. This motivates the following definition.

Definition. Let the density $\rho(t, \underline{x})$, $\underline{x} \in \Omega$, be a positive (C^1) function and let $\underline{v}$ be a vector field with flow $\phi(t, \underline{x})$. We say $\rho, \underline{v}$ satisfy the principle of conservation of mass if

$$\frac{d}{dt} \int_{\phi_t(W)} \rho(t, \underline{x})dx = 0$$

for every (nice) subregion W of Ω.

Theorem. *The principle of conservation of mass is satisfied by* $\rho, \underline{v}$ *if and only if any of the following equivalent conditions hold:*

(i) $\frac{D\rho}{Dt} + \rho \operatorname{div} \underline{v} = 0$,

(ii) $\frac{\partial\rho}{\partial t} + \operatorname{div}(\rho\underline{v}) = 0$,

(iii) $$\frac{d}{dt}\int_W \rho\, dx = -\int_S \rho\underline{v},$$

for any nice subregion W *of* Ω *and* $S = \partial W$.

Equations (i) and (ii) are both referred to as the equation of continuity.

Proof. That the definition and (i) are equivalent follows from the transport theorem with $f = \rho$; (i) and (ii) are equivalent is deduced from the definition of material derivative and the identity $\operatorname{div}(\rho\underline{v}) = \nabla\rho\cdot\underline{v} + \rho \operatorname{div} \underline{v}$. In the left hand side of (iii) we can differentiate under the integral sign since W is fixed and ρ is C^1. Applying Gauss' Theorem to the right-hand side of (iii) shows then that (iii) is equivalent to (ii). ∎

Corollary. *Suppose* ρ, $\underline{v}$ *satisfy the principle of conservation of mass. Then the flow of* $\underline{v}$ *is incompressible if and only if* $D\rho/Dt = 0$.

Proof. Recall incompressible is equivalent to $\operatorname{div} \underline{v} = 0$; the conclusion follows from (i). ∎

Theorem. *Let* $\rho, f\colon \mathbb{R} \times \Omega \to \mathbb{R}$ *and let* $\underline{v}$ *be a vector field with flow* $\underline{\phi}(t, \underline{x})$. *Then if* ρ, $\underline{v}$ *satisfy the principle of conservation of mass,*

$$\frac{d}{dt} \int_{\underline{\phi}_t(W)} \rho f \, dx = \int_{\underline{\phi}_t(W)} \rho \frac{Df}{Dt} \, dx$$

for all nice subregions $W \subset \Omega$.

Proof. In the transport theorem replace f by ρf and use the continuity equation; $\frac{D(\rho f)}{Dt} = \rho \frac{Df}{Dt} + (\frac{D\rho}{Dt}) f = \rho \frac{Df}{Dt} - \rho \nabla \cdot \underline{v} f$. ∎

If ρ is constant in $\underline{x}$ we call the fluid homogeneous. If the fluid is homogeneous and incompressible then from the above, ρ is constant in t as well.

Problem. Derive a formula akin to the transport theorem for

$$\frac{d}{dt} \int_{\underline{\phi}_t(S)} \underline{w} = \int_{\phi_t(S)} \left[\frac{\partial w}{\partial t} + v(\nabla \cdot w)\right]$$

where $\underline{\phi}_t(S)$ is a moving closed surface and $\underline{w}$ is a vector field.

Problem. Show that another way to write the equation of continuity is

$$\rho(t, \underline{\phi}(t, \underline{x})) J(t, \underline{x}) = \rho(0, \underline{x}) .$$

§7. Balance of Momentum; Cauchy's Stress Principle.

We quote C. Truesdell* concerning the Stress Principle of Cauchy: "upon any smooth, closed, orientable surface $\mathscr{S}$, be it an imagined surface within the body or the bounding surface of the body itself, there exists an integrable field of traction $\underline{t}_{\mathscr{S}}$ equipollent (same resultant and moment) to the action exerted by the matter exterior to $\mathscr{S}$ and contiguous to it on that interior of $\mathscr{S}$."

This basic idea leads one to (see Figure 7-1):

Definition. Let $\underline{t}(t, \underline{x}, \underline{n})$ be a given vector field depending on t, $\underline{x}$ and a unit vector $\underline{n}$. Let $\underline{v}$ and ρ be a vector field and a positive function and $\underline{f}$ a given vector field. We say $(\underline{v}, \rho, \underline{f}, \underline{t})$ satisfy the balance of momentum principle if

$$\frac{d}{dt} \int_{\underline{\phi}_t(W)} \rho \underline{v} \, dx = \int_{\underline{\phi}_t(W)} \rho \, \underline{f} \, dx + \int_{S_t} \underline{t} \, da \qquad (1)$$

where S_t is the boundary of $\underline{\phi}_t(W)$ and da is the area form for S_t . Note that this is a vector equation.

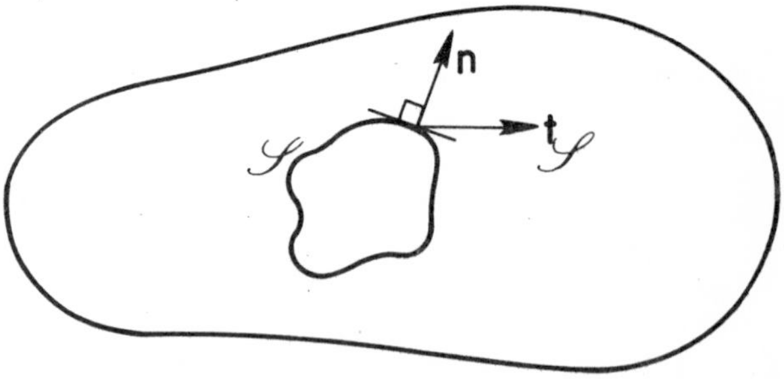

FIG. 7-1

*C. Truesdell, "Essays in the History of Mechanics" Springer (1968) p. 186. This principle expresses the idea of a continuous material acting within itself by surface contact forces.

Here $\underline{v}$ represents the velocity field of a fluid, ρ the mass density, $\underline{f}$ the external force, and $\underline{t}\,da$ the internal forces acting on a surface with area element da oriented with normal $\underline{n}$.

Cauchy's Theorem. *The balance of momentum implies that there exists a matrix function* $T_{ij}(t, \underline{x})$ *such that* $\underline{t}(t, \underline{x}, \underline{n}) = \underline{T}(t, \underline{x})\underline{n}$.

Proof. Fix t at t_0 and assume $W_{t_0} = \underline{\phi}_{t_0}(W)$ is a tetrahedron with three faces aligned with the coordinate planes such that the unit normal vector $\underline{n}$ to the skew face satisfies $n_1 > 0$, $n_2 > 0$, $n_3 > 0$ (see Figure 7-2). Take the volume of W_{t_0} to be ℓ^3, where ℓ has di-

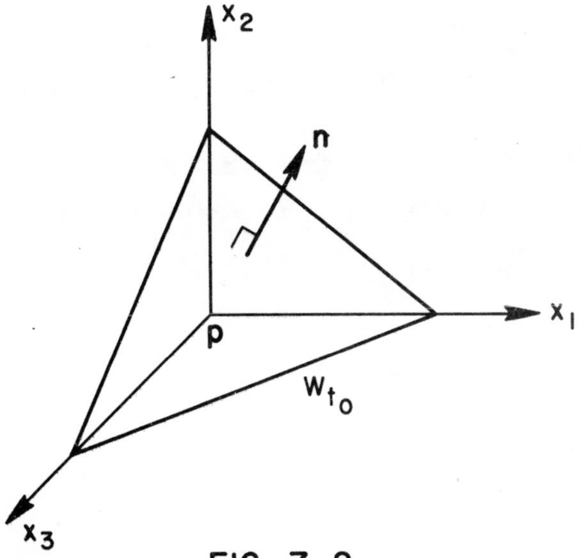

FIG. 7-2

mensions of length. The balance of momentum may be written

$$\int_{S_{t_0}} \underline{t}\,da = \int_{W_{t_0}} \left(\rho \frac{D}{Dt}\underline{v} + (D\rho/Dt + \rho\,\mathrm{div}\,\underline{v})\underline{v} - \rho\underline{f}\right)dx ,$$

where S_{t_0} is the surface of W_{t_0}, by the Transport Theorem. Assuming the integrand of the right hand side is bounded by $M = \text{constant} > 0$ throughout W_{t_0},

$$\| \int_{S_{t_0}} \underline{t}\, da \| \leqslant \ell^3 M .$$

The area of S_{t_0} is equal to a constant times ℓ^2. Thus in the limit as the volume of W_{t_0} approaches zero,

$$\lim_{\ell \to 0} \frac{1}{\ell^2} \int_{S_{t_0}} \underline{t}\, da = 0 .$$

Denote the face perpendicular to the x_i direction by $\sum_i$ and the skew face by $\sum$. The area of $\sum$ is equal to a positive constant, say c, times ℓ^2. Thus by elementary geometric considerations, the area of $\sum_i$ is $n_i c \ell^2$. By the continuity of $\underline{t}$, the mean value theorem gives

$$\frac{1}{\ell^2} \int_{S_{t_0}} \underline{t}\, da = c\{\underline{t}(t_0, \underline{z}, \underline{n}) + n_1 \underline{t}(t_0, \underline{z}_1, -\underline{i}) + n_2 \underline{t}(t_0, \underline{z}_2, -\underline{j}) + n_3 \underline{t}(t_0, \underline{z}_3, -\underline{k})\} \qquad (2)$$

for some points $\underline{z}$ in $\sum$ and $\underline{z}_i$ in $\sum_i$. Let the volume of the tetrahedron approach zero by letting the skew face approach p, the vertex opposite the skew face. The left hand side of (2) vanishes in this limit. Therefore, dividing through by c,

$$0 = \underline{t}(t_0, p, \underline{n}) + n_1\underline{t}(t_0, p, -\underline{i})$$
$$+ n_2\underline{t}(t_0, p, -\underline{j}) + n_3\underline{t}(t_0, p, -\underline{k}) .$$

We can omit the dependence on t_0 and p for simplicity, i.e.,

$$0 = \underline{t}(\underline{n}) + n_1\underline{t}(-\underline{i}) + n_2\underline{t}(-\underline{j}) + n_3\underline{t}(-\underline{k}) . \qquad (3)$$

Here, let $n_1 \to 1$, n_2 and $n_3 \to 0$; then $\underline{t}(\underline{i}) = -\underline{t}(-i)$. Similarly $\underline{t}(\underline{j}) = -\underline{t}(-\underline{j})$ and $\underline{t}(\underline{k}) = -\underline{t}(-\underline{k})$. Plugging these in (3) yields

$$\underline{t}(\underline{n}) = n_1\underline{t}(\underline{i}) + n_2\underline{t}(\underline{j}) + n_3\underline{t}(\underline{k}) .$$

The same result can be obtained if $\sum$ is located in any other octant. Therefore $\underline{t}$ is linear in $\underline{n}$. Since it linearly transforms vectors $\underline{n}$ into vectors $\underline{t}(\underline{n})$, we conclude that there exists a matrix $\underline{T}(t, \underline{x})$ such that $\underline{t}(t, \underline{x}, \underline{n}) = \underline{T}(t, \underline{x})\cdot\underline{n}$. ■

$\underline{T}$ is in fact a tensor and is called the <u>stress tensor</u> (this fact follows from the "quotient rule" of tensor analysis).

<u>Problem</u>. Show that balance of momentum can be rephrased as

$$\frac{d}{dt}\int_W \rho\underline{v}\, dx = \int_W \rho\underline{f}\, dx + \int_S (\underline{t} - \rho\underline{v}(\underline{v}\cdot\underline{n}))da .$$

(Here W is <u>fixed</u>.) Interpret this formula physically.

§8. Balance of Momentum; Equations of Motion.

We now deduce local forms for the balance of momentum. The balance of momentum, in component form, is

$$\frac{d}{dt}\int_{W_t} \rho v_i \, dx = \int_{W_t} \rho f_i \, dx + \int_{S_t} T_{ij} n_j \, da \ . \tag{1}$$

Employing the notation $(\underline{T}_i)_j = T_{ij}$, Gauss' Theorem and the Transport Theorem result in

$$0 = \int_{W_t} \left(\frac{D}{Dt}(\rho v_i) + \rho v_i \operatorname{div} \underline{v} - \rho f_i - \operatorname{div} \underline{T}_i \right) dx \ , \tag{2}$$

where $\operatorname{div} \underline{T}_i = \partial T_{ij}/\partial x_j$, with summation on j understood. By continuity of the integrand, we can localize (2); it holding for all W is the same as

$$\frac{D}{Dt}(\rho v_i) + \rho v_i \operatorname{div} v = \rho f_i + \operatorname{div} \underline{T}_i \ , \tag{3}$$

which is equivalent to

$$\boxed{\rho \frac{Dv_i}{Dt} = \rho f_i + \operatorname{div} \underline{T}_i} \ , \tag{4}$$

if the continuity equation holds. (4) is known variously as the local form of the momentum balance, Cauchy's first law and the equation of motion.

The equations (4) together with the equation of continuity (§6) govern how the fluid moves in time, i.e., how $\underline{v}$ and ρ behave as functions of t. However, the description is not complete unless we know the stress tensor $\underline{T}$, or have an equation for it, and have suitable boundary conditions specified. These points will all be considered in the following sections.

Essay Topic: Discuss the derivation of the equations of motion from statistical mechanics. See Shinbrot [13].

Balance of Angular Momentum. Let ρ, $\underline{v}$, $\underline{f}$, $\underline{t}$, etc. be as before. We say the balance of angular momentum is satisfied if

$$\frac{d}{dt}\int_{W_t} \rho(\underline{x} \times \underline{v})dx = \int_{W_t} \rho(\underline{x} \times \underline{f})dx + \int_{S_t} \underline{x} \times \underline{t}\, da ,$$

where $\underline{x}$ is the position vector and $W_t = \phi_t(W)$. Note that the left hand side is the rate of change of angular momentum.

Cauchy's Second Law. Assume balance of momentum and the equation of continuity are satisfied. Then balance of angular momentum holds if and only if $T_{ij} = T_{ji}$, i.e., the stress tensor is symmetric.

Problem. Prove Cauchy's second law (use the same methods as in the previous theorem).

§9. Perfect Fluids and Euler's Equations.

Definition. A fluid is said to be perfect if it can exert no tangential stresses on a surface, i.e., $\underline{T}\cdot\underline{n}$ is parallel to $\underline{n}$ for all unit vectors $\underline{n}$.

By the algebraic lemma (§3), this implies that $T_{ij} = -p\delta_{ij}$ where p is a scalar valued function. It follows that $\underline{T}\cdot\underline{n} = -p\underline{n}$ is the force/unit area on a surface with normal $\underline{n}$.

The equations of motion for a perfect fluid are:

$$\rho \frac{D\underline{v}}{Dt} = \rho\underline{f} - \underline{\nabla}p \ , \quad \text{i.e.} \ , \quad \rho \frac{Dv_i}{Dt} = \rho f_i - \frac{\partial p}{\partial x_i} \ , \tag{1}$$

since it is easily verified that, in this case, $\operatorname{div} \underline{T} = -\underline{\nabla}p$.

Definition. An ideal fluid is an incompressible perfect fluid.*

The field equations for an ideal fluid are (Euler's Equations):

$$\left.\begin{aligned} \rho \frac{D\underline{v}}{Dt} &= \rho\underline{f} - \underline{\nabla}p \\ \frac{D\rho}{Dt} &= 0 \\ \operatorname{div} \underline{v} &= 0 \end{aligned}\right\} \tag{2}$$

The first equation is the same as $\rho \frac{\partial\underline{v}}{\partial t} + \rho(\underline{v}\cdot\underline{\nabla})\underline{v} = \rho\underline{f} - \underline{\nabla}p$. This system of equations is (formally) deterministic, i.e., five equations in five unknowns $(\rho, \underline{v}, p)$. The Cauchy, or initial-value problem,

*Often ρ constant in $\underline{x}$ is taken to be part of the definition of an ideal fluid, but this need not be the case here.

consists of solving (2) on all of $\mathbb{R}^2$ or $\mathbb{R}^3$ given ρ, $\underline{v}$ at $t = 0$. A boundary value problem consists of (2) on Ω, a region with boundary $\partial\Omega$, ρ, $\underline{v}$ given at $t = 0$ and certain boundary data specified on $\partial\Omega$. For example, if Ω is bounded by rigid walls, $\underline{v} \parallel \partial\Omega$ is the appropriate boundary condition. (More about this later.)

Existence Theorem.* *If the initial conditions* ρ_0, $\underline{v}_0$ *are sufficiently smooth there exists a unique solution* ρ, $\underline{v}$, p (p *up to an additive constant*) *to* (2) *for the Cauchy problem with the rigid boundary conditions* $\underline{v} \parallel \partial\Omega$ ($\partial\Omega$ *sufficiently smooth*). *In general, the solution only exists for a short time.*

Theorem† (Wolibner, 1932). *If* $n = 2$ (*two dimensions*) *and* $\rho = 1$ *solutions as above exist and remain smooth for all* t.

A main open problem in the theory of ideal fluids is to deduce necessary and sufficient conditions on initial data for a solution to exist for all t in three dimensions.

---

*See L. Lichtenstein "Grundlagen der Hydromechanik," Springer (1929), D. Ebin and J. Marsden, Ann. of Math. 92(1970) 102-163, T. Kato, J. Funct. An. 9(1972) 296-305, J. Bourguignon and H. Brezis, J. Funct. An. 12(1973) and R. Temam, J. Funct. An. 20(1975) 32-43. The case of variable ρ is found in J. Marsden (unpublished). The smoothness is in class $C^{1+\alpha}$ or $W^{s,p}$.

†See W. Wolibner, Math. Zeit. 37(1933) 698-726, V. Judovich, Mat. Sb.N.S. 64(1964) 562-588 and T. Kato, Arch. Rat. Mech. An. 25(1967) 188-200.

A <u>stationary solution</u> of the Euler equations is a triple ρ, $\underline{v}$, p, independent of t such that

$$\left.\begin{aligned} \rho(\underline{v}\cdot\underline{\nabla})\underline{v} &= \rho\underline{f} - \underline{\nabla}p \\ \underline{\nabla}\cdot(\rho\underline{v}) &= 0 \\ \text{and} \qquad \operatorname{div}\underline{v} &= 0 \end{aligned}\right\} \tag{3}$$

This is the same as (2) with $\partial/\partial t$ terms omitted. We similarly define what we mean by a stationary solution of (1).

<u>Problem</u>. Prove the identity:

$$(\underline{v}\cdot\underline{\nabla})\underline{v} = \frac{1}{2}\underline{\nabla}(v^2) + (\underline{\nabla}\times\underline{v})\times\underline{v} \tag{4}$$

where $v = \|\underline{v}\| = (v_1^2 + v_2^2 + v_3^2)^{1/2}$.

§10. Potential Flows.

Definition. A flow is potential if the velocity field satisfies $\underline{v} = \underline{\nabla}\phi$, where ϕ is a real valued function (called the potential function).

A flow is irrotational if $\underline{\nabla} \times \underline{v} = \underline{0}$.

It is immediate that potential flows are irrotational. The converse is not necessarily true. If $\underline{v}$ is potential and incompressible then ϕ satisfies Laplace's equation $\Delta\phi = 0$ $\left(\Delta = \frac{\partial^2}{\partial x^2} + \frac{\partial^2}{\partial y^2} + \frac{\partial^2}{\partial z^2}\right)$.

Example. (An irrotational flow which is not a potential flow.) Consider the irrotational flow

$$\underline{v} = \frac{-y}{x^2+y^2}\,\underline{i} + \frac{x}{x^2+y^2}\,\underline{j} \quad \text{on } \mathbb{R}^2 - \{\underline{0}\} ;$$

one sees $\underline{\nabla} \times \underline{v} = 0$. However, there does not exist a ϕ such that $\underline{\nabla}\phi = \underline{v}$. To see this, suppose the contrary was true, i.e., ϕ exists such that $\underline{\nabla}\phi = \underline{v}$. Integrate both sides about a circle containing the origin (Figure 10-1):

$$\int_C \underline{\nabla}\phi\cdot\underline{ds} = 0 ,$$

since C is closed (see §3). On the other hand, by direct calculation,

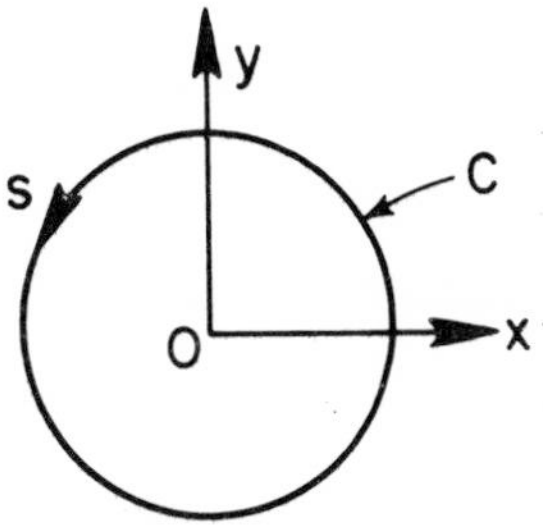

FIG. 10-1

$$\int_C \underline{v} \cdot \underline{da} = -2\pi \ .$$

Contradiction! Therefore the original hypothesis was incorrect, i.e., there does not exist any ϕ such that $\underline{\nabla}\phi = \underline{v}$.

The reason why this occurs is that $\mathbb{R}^2 - \{0\}$ is not simply connected.* For simply connected regions $\underline{\nabla} \times \underline{v} = \underline{0} \Leftrightarrow$ there exists ϕ such that $\underline{\nabla}\phi = \underline{v}$. Thus locally we can always find a ϕ for an irrotational flow, i.e., in each simply connected subregion we can find ϕ . The example illustrates that globally this is not possible.

Bernoulli's Law. Let $\underline{v}$ be a potential flow which satisfies the equations of motion ((1) of §9) and assume that $\underline{\mathbf{f}} = -\underline{\nabla}K$, where K is a scalar valued function. Consider a path C connecting two points a and b and assume C is contained in the region where $\underline{v} = \underline{\nabla}\phi$. Then

$$\left(\frac{\partial\phi}{\partial t} + \frac{1}{2} v^2 + K \right) \Big|_a^b + \int_C \frac{dp}{\rho} = 0 \ .$$

Proof. Integrate (1) §9, using (4), over the path C . ■

Special Case. (1) If dp/ρ is an exact differential, i.e., $dp/\rho = d\tilde{p}$ then $\int_C \frac{dp}{\rho} = \int_C d\tilde{p} = \tilde{p}\Big|_a^b$. For example, if ρ = constant, $\tilde{p} = p/\rho$.

---

*A region is called simply connected if any closed loop can be shrunk continuously down to a point in the region, but never leaving the region.

(2) For a stationary flow (i.e., $\underline{v}(t, \underline{x}) = \underline{v}(\underline{x})$) which is irrotational, incompressible and ρ = constant,

$$\frac{1}{2} v^2 + \frac{1}{\rho} p + K = \text{constant in the region.}$$

Note that it is not necessary for $\underline{v}$ to be potential for (2) to hold. One needs only that $\underline{v}$ is stationary, imcompressible and irrotational. Case (2) is the origin of the ubiquitous: "where the velocity is higher the pressure is lower."

Theorem. *Let* $\underline{v}$ *be stationary, incompressible and irrotational and assume* ρ *is constant. Then* $\underline{v}$ *is a solution to Euler's equations* (§9) *in which* $p = \frac{-\rho}{2} v^2$ *and* $\underline{f} = 0$.

Proof.
$$-\underline{\nabla} p = \rho \frac{D\underline{v}}{Dt} = \rho(\underbrace{\frac{\partial \underline{v}}{\partial t}}_{0} + (\underline{v}\cdot\underline{\nabla})\underline{v}) = \rho\{\underbrace{(\underline{\nabla} \times \underline{v})}_{0} \times \underline{v} + \frac{1}{2} \underline{\nabla}(v^2)\}$$
$$= \underline{\nabla}(\frac{\rho}{2} v^2) .$$ ■

Remarks. (1) Irrotational can be replaced by potential in the above. This is the way one is led to study harmonic functions (i.e., solutions of $\Delta\phi = 0$) in fluid mechanics, since $\underline{v} = \underline{\nabla}\phi$ are stationary solutions of Euler's equations when ρ = constant . This has the nice feature of reducing the nonlinear Euler equations to a linear problem which is well understood.

(2) A stationary body force stemming from a potential, $\underline{f} = -\underline{\nabla}K$, can be included by changing p to $p + K$.

Problem. Let ρ, $\underline{v}$ be a stationary solution of the equations of motion (1, §9), not necessarily incompressible nor a potential or irrotational flow. Show that along any given particle path (= streamline), say $c(s)$,

$$\frac{1}{2} v^2(c(s)) + K(c(s)) + \int_{c(s_0)}^{c(s)} \frac{dp}{\rho}$$

is constant in s (the integral is along the path $c(s)$ from a fixed end $c(s_0)$ out to a variable end $c(s)$, and $\underline{f} = -\underline{\nabla} K$). Hint. Use (4) of §9 and note that $(\underline{\nabla} \times \underline{v}) \times \underline{v}$ is orthogonal to $c(s)$.

§11. The Kutta-Joukowski Theorem and the Paradox of Lift.

Consider flow over an airplane wing (Figure 11-1). Since the velocity is higher on the top (the fluid has to go farther), it is tempting to say that the pressure is lower because of Bernouli's law and hence results in lift.

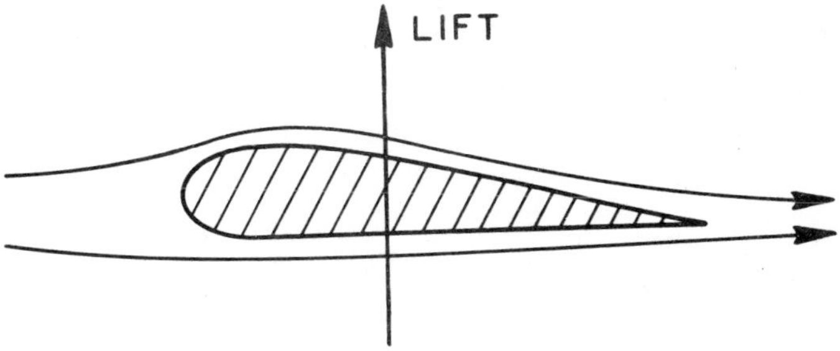

FIG. 11-1

Unfortunately, this is very naive. In the first place, it is not clear at all that the hypotheses of Bernouli's law (§10) are relevant here. In reality one has to take the viscous nature of the fluid and the boundary conditions into account much more seriously.

In fact we shall prove below (and later in §13) that lift can come about in ideal flow only under very artificial circumstances* in $\mathbb{R}^2$ and not at all in $\mathbb{R}^3$!

Thus the paradox of lift is not really a paradox; it is just the fact that the model of an ideal fluid is inadequate to explain the real-life fact of lift.

---

*Various ad hoc assumptions (e.g.,"Joukowski's hypothesis") are usually added which in fact try to simulate non-negligible viscous effects.

There are two more "paradoxes" we would like to mention, while we are on the subject. Both are given to illustrate the caution needed when applying general principles or intuitive arguments to complicated situations.

First, consider the "Magnus effect." This is the effect that enables pitchers and tennis players (good ones) to deliver curve balls. When a rapidly rotating ball is thrown in a fluid it drifts as shown in Figure 11-2.

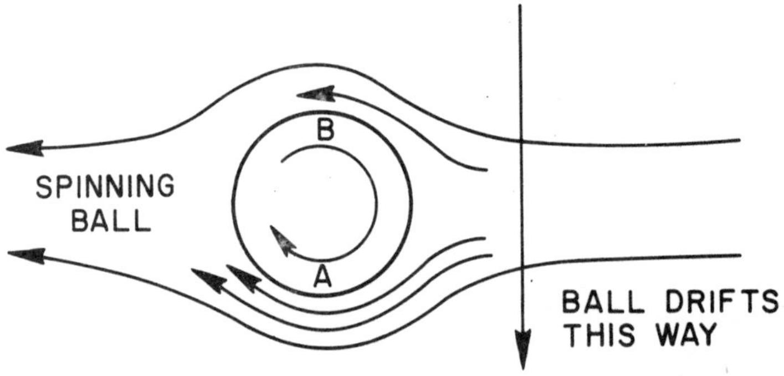

FIG. 11-2

Usual explanation: Because the ball is spinning, the fluid is swept around faster on side A than on B. Hence the pressure is lower there, so the ball drifts as shown.

Real-life fact: If the ball is spinning slowly, the ball drifts the other way!

Paradox! In reality, the motion of the fluid, especially near the boundary is complicated; certainly not irrotational or stationary. Bernoulli's results are simply not relevant. Thus there is no paradox, but there is no simple explanation either.

Another amusing paradox is the "Dubat paradox." It says this: dragging a stick through still water produces more resistance than moving water (e.g., a stream) flowing past a stick with the same velocity. This seems to contradict common sense. In fact, flow in a stream is usually turbulent or chaotic and it has quite different properties from steady more regular flow; it is observed quite generally that turbulence lowers the resistence. However, a precise explanation of why this is so is not known.

Let us return now and investigate two-dimensional irrotational, incompressible flow in more detail, using complex variable methods. (See J. Marsden, "Basic Complex Analysis," Freeman (1973), for all the background needed and for additional problems.)

Let us begin by recalling the following basic result:

Theorem. *Let* $F = u + iv$, *where* u, v *are* C^1 *functions from an open set* $U \subset \mathbb{R}^2$ *into* $\mathbb{R}$. *In addition, let* u, v *satisfy*

$$\frac{\partial u}{\partial x} = \frac{\partial v}{\partial y} \quad \text{and} \quad \frac{\partial u}{\partial y} = \frac{-\partial v}{\partial x}$$

(the Cauchy-Riemann equations). Then F *is an analytic function on* U , *considered as an open set in the complex plane* $\mathbb{C}$.

<u>Corollary</u>. *Consider a two-dimensional flow with velocity field* $\underline{v} = v_x \underline{i} + v_y \underline{j}$. *If the flow is incompressible and irrotational, i.e.,*

$$\left.\begin{aligned} \underline{\nabla} \cdot \underline{v} = 0 &\Leftrightarrow \frac{\partial v_x}{\partial x} + \frac{\partial v_y}{\partial y} = 0 \\ \underline{\nabla} \times \underline{v} = \underline{0} &\Leftrightarrow \frac{\partial v_x}{\partial y} - \frac{\partial v_y}{\partial x} = 0 \end{aligned}\right\} \qquad (1)$$

then $F(z) = v_x - iv_y$, *called the <u>complex velocity</u>, is an analytic function of* $z = x + iy$. *Conversely, if* F *is analytic then* (1) *holds. In particular, if the flow is incompressible and potential, then* F *is analytic.*

If F is analytic on a simply connected domain then $F = df/dz$, where f is an analytic function. In this case, f is called the <u>complex potential</u> and ϕ is the real part of f . Note that in general, an analytic function is not the derivative of another analytic function. We emphasize this for in the ensuing applications we will not need the existence of a complex potential. We will only need that F is analytic.

<u>Blazius Theorem</u>. *Let* C *be a simple closed curve in* $\mathbb{R}^2$ *which is smooth and let* $\underline{v}$ *be a stationary two-dimensional velocity field defined on the exterior of* C *such that* $\underline{v}$ *is parallel to* C *on* C . *(See Figure* 11-3.) *In addition, assume* $\underline{v}$ *is an incompressible, irrotational, Euler flow corresponding to* ρ = *constant. Define the* x *and* y *forces on*

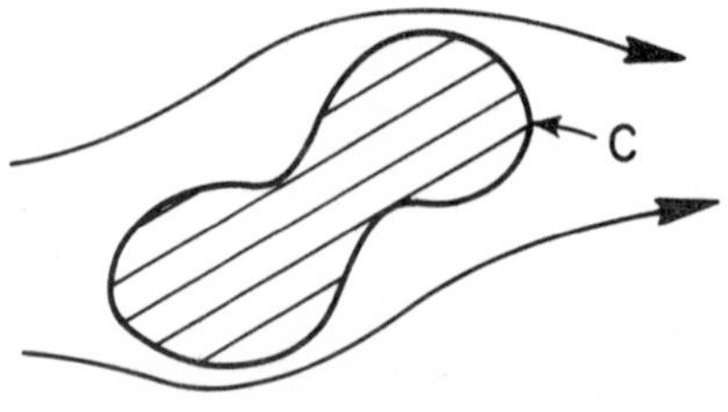

FIG. 11-3

the "obstacle" bounded by C *to be*‡

$$F_x = -\int_C p\, dy ,$$

$$F_y = \int_C p\, dx ,$$

respectively, where p *is the pressure. Then*

$$F_x - iF_y = \frac{i\rho}{2}\int_C F^2\, dz ,$$

where $F = v_x - iv_y$.

Proof. First we obtain an expression for the left-hand side:

$$\begin{aligned} F_x - iF_y &= -\int_C p(dx - idy) \\ &= \frac{\rho}{2}\int_C v^2(dx - idy) \\ &= \frac{\rho}{2}\int_C (v_x^2 + v_y^2)(dx - idy) . \end{aligned}$$

Line two follows from line one by a previous theorem on Euler flows (p. 39).

The right-hand side,

---

‡This "definition" emanates from the traction vector expression for a perfect fluid; $\underline{t} = -p\underline{n}$.

$$\frac{i\rho}{2}\int_C F^2 dz = \frac{i\rho}{2}\int_C (v_x - iv_y)^2(dx + idy) ,$$

when combined with the differential representation of the condition $\underline{v}$ parallel to C (i.e., $v_x dy = v_y dx$) , is identical to the left-hand side (verify!). ■

Note that $\underline{v}$ parallel to a boundary implies $F dz = \underline{v}\cdot\underline{ds}$, and conversely.

Example: Let $F = U =$ constant on the upper half-plane (see Figure 11-4).

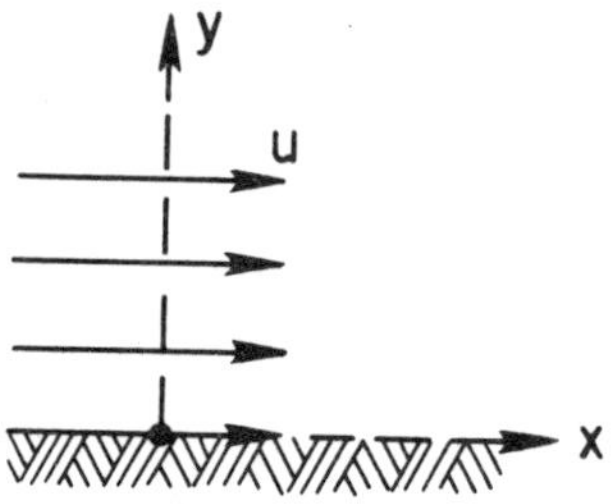

FIG. 11-4

Then F is analytic and $\underline{v}$ is incompressible and irrotational.

We can convert a complex velocity on one region to one on another, by using the following.

Riemann Mapping Theorem. *Any two simply connected regions in the complex plane* $\mathbb{C}$ *(neither of which equals* $\mathbb{C}$ *) are conformally related, i.e., if* A, B *are simply connected regions in* $\mathbb{C}$, $\neq \mathbb{C}$, *then there exists an analytic function* $f: A \to B$ *which is one-to-one, onto and* $df/dz(z) \neq 0$ *for all* $z \in A$.

Problem. Define a mapping ζ from the upper half-plane minus the disc of radius a onto the upper half-plane (Figure 11-5) by $\zeta: z \to z + \frac{a^2}{z}$.

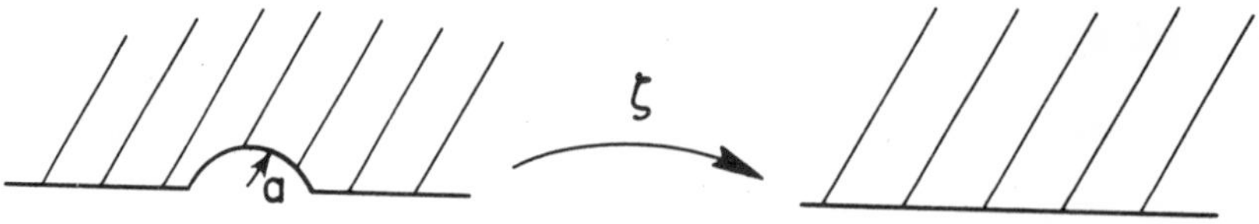

FIG. 11-5

This is known as a Joukowski transformation.

(a) Verify ζ is a conformal mapping and obtain the flow about a circular disc to be $F_1 = g'$ where $g = U(z + \frac{a^2}{z})$, U a constant. Show that there is no circulation about the disc.

(b) In $\mathbb{R}^2-\{\underline{0}\}$, show that $F_2(z) = i\Gamma/2\pi z$, Γ const., gives a circular flow, with circulation Γ on a circle. Conclude that the flow is irrotational and incompressible, but not potential. (Note that F_2 is analytic, but $g = i\Gamma/2\pi \log z$ is not analytic on $\mathbb{R}^2-\{0\}$; g is unnecessary in this problem.)

(c) Deduce that $F_1 + F_2$ is a flow around a disc of radius a , with circulation Γ and velocity U parallel to the x-axis at ∞ .

Kutta-Joukowski Theorem. *Assume the same hypotheses as in the Blazius theorem. In addition, let the velocity field become uniform* at* ∞ , *i.e., let* $v_x = U = const.$ *and* $v_y = 0$ *at* ∞ . *Then*

$$F_x = 0 \quad \text{and} \quad F_y = -\rho U\Gamma$$

where Γ *is the circulation around* C .

Proof. Since $\underline{v}$ is parallel to C , we have that

$$\int_C F\,dz = \int_C \underline{v}\cdot\underline{ds} \overset{\text{def.}}{=} \Gamma .$$

Now F is analytic, therefore it has a convergent Laurent expansion

---

*In a sense spelled out in the proof.

outside any disc:

$$F = B_1 z + B_2 z^2 + \dots + A_0 + \frac{A_1}{z} + \frac{A_2}{z^2} + \dots .$$

Since $(v_x, v_y) \to (U, 0)$ as $z \to \infty$, we conclude that $A_0 = U$ and the B_i's $\equiv 0$. By Cauchy's Theorem,

$$\int_C F dz = 2\pi i A_1 ,$$

thus $A_1 = \Gamma/2\pi i$, and

$$F = U + \frac{\Gamma}{2\pi i z} + \frac{A_2}{z^2} + \dots$$

We can compute F^2 by term-by-term multiplication (this is rigorous since F is analytic):

$$F^2 = U^2 + \frac{U\Gamma}{\pi i z} + \left(\frac{\Gamma}{2\pi i z}\right)^2 + \frac{2A_2 U}{z^2} + \dots$$

By Blazius' Theorem and Cauchy's Theorem (F^2 is analytic) we get

$$F_x - iF_y = i\rho U\Gamma . \qquad \blacksquare$$

<u>Remarks</u>. (1) F_x is referred to as the <u>drag</u> and F_y is the <u>lift</u>. The fact that there is no drag in the above situation is sometimes referred to as the "drag paradox."

(2) The contour C may be deformed into any simple closed curve surrounding C without changing anything, since F is analytic outside C.

(3) If the orientation of the uniform velocity field at ∞ is changed by a counter-clockwise motion through an angle α, then so will be the resultant forces.

As we shall see later in §14, the hypothesis that $\Gamma \neq 0$ is unreasonable if we stick entirely to the confines of ideal fluids. Also, as we all know, drag really does occur. One might suggest that restricting ourselves to stationary or irrotational or incompressible fluids is not realistic. There is some merit in these suggestions, but the real difficulty seems to be that we have neglected viscosity. However small, it has a non-negligible effect, especially near the boundary and cannot be ignored. This topic will be taken up in §16.

Problems. 1. Verify the Kutta-Joukowski theorem for the flow discussed in the previous problem (p. 46).

2. Let $f(z) = z^2$ be a complex potential in the first quadrant. Sketch some streamlines and the curves $\phi = \text{const.}$, $\psi = \text{const.}$, where $f = \phi + i\psi$. What is the force exerted on the walls?

§12. The Deformation Tensor and the Energy Transport Theorem.

Deformation Tensor. The symmetric part of the velocity gradient is known as the deformation tensor $\underline{D}$, i.e., $D_{ij} = \frac{1}{2}(v_{i,j} + v_{j,i})$ where we use the notations $, i = \frac{\partial}{\partial x_i}$ and $(x_1, x_2, x_3) = (x, y, z)$. This tensor has great significance in continuum mechanics and particularly in fluid mechanics. It is also geometrically interesting; for those who are familiar with basic differentiable manifold theory we note that $2\underline{D} = L_{\underline{v}}\underline{g}$, the Lie derivative of the metric tensor in the direction of the velocity vector, which measures the rate of change of local distances. Let us work this out in the present context:

To see how $\underline{D}$ measures local distance changes, let $c(s)$ be a curve in Ω, a region in $\mathbb{R}^3$. Under the motion $\underline{x} \mapsto \underline{\phi}_t(\underline{x})$, $\underline{c}(s) \mapsto \underline{\phi}_t(\underline{c}(s))$. The tangent vector $w = c'$ is transported to $D\underline{\phi}_t(\underline{c}(s))\cdot\underline{c}'(s)$, by the chain rule, from which it follows that any vector $\underline{w}$ at $\underline{x}$ ought to be defined to move under the flow by: $\underline{w}(t) = D\underline{\phi}_t(\underline{x})\cdot\underline{w}$ Let $\underline{w}_1, \underline{w}_2$ be vectors at $\underline{x}$ transported under the flow to $\underline{w}_1(t)$ and $\underline{w}_2(t)$ as just defined. Then

$$\begin{aligned}
\frac{d}{dt}\underline{w}_1(t)\cdot\underline{w}_2(t)\Big|_{t=0} &= \frac{d}{dt}(D\underline{\phi}_t(\underline{x})\cdot\underline{w}_1)\cdot(D\underline{\phi}_t(\underline{x})\cdot\underline{w}_2)\Big|_{t=0} \\
&= \frac{d}{dt}\left(\frac{\partial\phi^i}{\partial x^j} w_1^j \frac{\partial\phi^i}{\partial x^k} w_2^k\right)\Big|_{t=0} \qquad (\text{sum } i, j, k) \\
&= \frac{\partial v^i}{\partial x^j} w_1^j \delta_k^i w_2^k + \delta_j^i w_1^j \frac{\partial v^i}{\partial x^k} w_2^k \\
&= 2D_{ij} w_1^i w_2^j = 2\underline{w}_1^T \underline{D}\, \underline{w}_2 .
\end{aligned}$$

Here we write the component indices up rather than down to avoid confusion. Thus $\underline{w}_1 = (w_1^1, w_1^2, w_1^3)$, etc. Also, the summation convention will be in force from now on: summation on repeated indices is understood.

In line 3 we used the fact that $(D\underline{\phi}_0(\underline{x}))^i_j = \delta^i_j$, the Kronecker delta. Thus the time rate-of-change of the inner product of two time dependent vector fields is measured exclusively in terms of the deformation tensor. This tells us that the rates-of-change of angles and distances are determined by $\underline{D}$. We write

$$\frac{d}{dt}(\underline{w}_1(t)\cdot\underline{w}_2(t)) = 2\underline{w}_1^T(t)\underline{D}\,\underline{w}_2(t) \ . \tag{1}$$

It is interesting to consider a basis $\underline{e}_i$ transported by the flow. From (1) we have

$$\frac{1}{2}\frac{d}{dt}\underline{e}_i\cdot\underline{e}_j = D_{ij} \ .$$

Thus

$$\frac{1}{2}\frac{d}{dt}\|\underline{e}_i\|^2 = D_{ii} \ \text{(no sum)}$$

and

$$\frac{1}{2}\frac{d}{dt}\underline{e}_i\cdot\underline{e}_i \ \text{(sum)} = D_{ii} \ \text{(sum)} = \text{trace}(\underline{D}) = \text{div}\,\underline{v} \ .$$

The last line corresponds to local rate-of-change of volume. The deformation tensor can be further decomposed into the sum of a volume-preserving motion and a pure dilatation (= no shearing):

$$\underline{D} = (\text{div}\,\underline{v})\underline{I} + (\underline{D} - (\text{div}\,\underline{v})\underline{I})$$

Since the second term has zero trace, local volumes are preserved by it. That the first term preserves angles can be seen from line one above, i.e., $D_{ij} = 0$ for $i \neq j$.

Theorem. *Let* $\underline{v}$ *be a* C^2 *vector field on a region* $\Omega \subset \mathbb{R}^3$. *Then if* $\underline{x}$ *and* $\underline{x}_0$ *are points in* Ω

$$\underline{v}(\underline{x}) = \underline{v}(\underline{x}_0) + \underline{D}(\underline{x}_0)(\underline{x} - \underline{x}_0) + \underline{\omega}(\underline{x}_0) \times (\underline{x} - \underline{x}_0)/2 + \theta(\underline{x} - \underline{x}_0)$$

where $\underline{\omega} = \underline{\nabla} \times \underline{v}$ *and* $\theta(\underline{x} - \underline{x}_0)$ *is* $o(\|\underline{x} - \underline{x}_0\|)$, *i.e.*, $\theta(\underline{x} - \underline{x}_0)/\|\underline{x} - \underline{x}_0\| \to \underline{0}$ *as* $\|\underline{x} - \underline{x}_0\| \to 0$.

Proof. Since $\underline{v}$ is C^2 we may apply Taylor's formula:

$$\underline{v}(\underline{x}) = \underline{v}(\underline{x}_0) + \underline{\underline{\nabla v}}(\underline{x}_0)\cdot(\underline{x} - \underline{x}_0) + \theta(\underline{x} - \underline{x}_0)$$

where $\theta(\underline{x} - \underline{x}_0)$ is $o(\|\underline{x} - \underline{x}_0\|)$. In coordinates, $(\underline{\nabla v})_{ij} = v_{i,j} = D_{ij} + \Omega_{ij}$, where $2D_{ij} = v_{i,j} + v_{j,i}$ and $2\Omega_{ij} = v_{i,j} - v_{j,i}$; $\underline{\Omega}$ is the spin tensor. It is easy to verify that $-\underline{\omega} \times \underline{r} = 2\underline{r}\cdot\underline{\Omega}$; in coordinates, $-\varepsilon_{ijk}\omega_j r_k = 2r_j\Omega_{ji}$, where ε_{ijk} is the alternator, defined by

$$\varepsilon_{ijk} = \begin{cases} 1 & \text{if } ijk \text{ is an even permutation of } 123 \\ -1 & \text{if } ijk \text{ is an odd permutation of } 123 \\ 0 & \text{otherwise} \end{cases}$$

(the student should verify these manipulations). ■

Remark. This result is given the following physical interpretation. One says that the velocity is, to first order, composed of a rigid translation, $\underline{v}(\underline{x}_0)$, a rigid rotation, $\frac{1}{2}\underline{\omega}(\underline{x}_0) \times (\underline{x} - \underline{x}_0)$, and a stretching and shearing, namely $\underline{D}(\underline{x}_0)\cdot(\underline{x} - \underline{x}_0)$.

Equation of Energy Transfer.

The kinetic energy $K(W_t)$ of a moving volume $W_t = \underline{\phi}_t(W)$ of fluid is given by the formula:

$$K = K(W_t) = \frac{1}{2}\int_{W_t} \rho v^2 \, dx \, .$$

Theorem (Energy Transport Theorem). *Assume that the equations of motion and continuity hold, and that the stress tensor* $\underline{T}$ *is symmetric. Then the rate-of-change of kinetic energy is given by:*

$$\frac{d}{dt} K = \int_{W_t} \rho \underline{f}\cdot\underline{v} \, dx + \int_{S_t} \underline{t}\cdot\underline{v} \, da - \int_{W_t} \underline{T}\cdot\underline{D} \, dx$$

where $S_t = \partial \, \underline{\phi}_t(W)$ *and* $\underline{T}\cdot\underline{D} = T_{ij}D_{ij}(\text{sum})$.

Proof. By a corollary of the transport theorem we have

$$\begin{aligned} \frac{d}{dt} K &= \frac{1}{2}\frac{d}{dt}\int_{W_t} \rho v^2 \, dx \\ &= \frac{1}{2}\int_{W_t} \rho \frac{D}{Dt}(v^2) \, dx \\ &= \int_{W_t} \rho\left(\frac{D}{Dt}\underline{v}\right)\cdot\underline{v} \, dx \, . \end{aligned}$$

The last line follows from the product rule applied component-wise: $D/Dt\,(v_i \cdot v_i) = 2v_i Dv_i/Dt$. Employing the equation of motion in the above, integrating by parts and using Gauss' theorem yields

$$\int_{W_t} \rho\left(\frac{D}{Dt}\underline{v}\right)\cdot\underline{v}\,dx = \int_{W_t} (\rho\underline{f} + \operatorname{div}\underline{T})\cdot\underline{v}\,dx$$

$$= \int_{W_t} (\rho\underline{f}\cdot\underline{v} + T_{ij,j}v_i)dx$$

$$= \int_{W_t} (\rho\underline{f}\cdot\underline{v} - T_{ij}v_{i,j})dx + \int_{S_t} T_{ij}n_jv_i\,da$$

$$= \int_{W_t} (\rho\underline{f}\cdot\underline{v} - \underline{T}\cdot\underline{D})dx + \int_{S_t} \underline{t}\cdot\underline{v}\,da\ .$$

The last line follows from the symmetry of $\underline{T}$ and the definition of the traction vector $\underline{t}$. ■

Corollary. *Suppose* $W_t = \Omega$, *a fixed region, and that* $\underline{v}$ *is parallel to* $\partial\Omega$. *Assume* $\underline{f} = 0$ *and the fluid is perfect and incompressible. Then* $K(\Omega)$ *is constant.*

Proof. For a perfect fluid (recall $T_{ij} = -p\delta_{ij}$) we have $\underline{T}\cdot\underline{D}$ $= (-p\delta_{ij})D_{ij} = -pD_{ii} = -p \operatorname{div}\underline{v} = 0$; by incompressibility. The traction vector $t_i = T_{ij}n_j = -p\delta_{ij}n_j = -pn_i$, i.e., it points in the direction of the outward unit normal vector $\underline{n}$. Thus $\underline{t}\cdot\underline{v} = 0$ since $\underline{v}$ is perpendicular to $\underline{n}$. ■

Remark. If the external body force emanates from a potential, i.e., if $\underline{f} = -\operatorname{grad}\phi$ for some function $\phi: \Omega \to \mathbb{R}$, then the corollary can be modified to read $K(\Omega) + V(\Omega)$ is constant, where $V(\Omega) = \int_\Omega \rho\,\phi\,dx$.

To see this note that

$$\frac{d}{dt} V = \int_\Omega \rho \frac{D}{Dt} \phi \, dx$$

$$= \int_\Omega \rho (\mathrm{grad}\ \phi) \cdot \underline{v} \, dx ,$$

by the chain rule.

Problem (Couette Flow). Let Ω be the region between two concentric cylinders of radii R_1 and R_2, $R_1 < R_2$. Define $\underline{v}$ in cylindrical coordinates by

$$v_r = 0 , \quad v_z = 0$$

$$v_\theta = \frac{A}{r} + Br$$

where

$$A = \frac{R_1^2 R_2^2 (\omega_2 - \omega_1)}{R_2^2 - R_1^2}$$

$$B = \frac{R_1^2 \omega_1 - R_2^2 \omega_2}{R_2^2 - R_1^2} .$$

Show (a) $\underline{v}$ is a stationary solution of Euler's equations with $\rho \equiv 1$.

(b) $\underline{\omega} = \underline{\nabla} \times \underline{v} = (0, 0, 2B)$.

(c) $\underline{D}$ (the deformation tensor) is $\frac{-A}{r^2} \begin{pmatrix} 0 & 1 \\ 1 & 0 \end{pmatrix}$.

Discuss the physical meaning of this result.

(d) The angular velocity of the flow on the two cylinders is ω_1 and ω_2.

§13. Momentum Transfer and d'Alembert's Paradox.

Theorem (Momentum Transfer). *Let* W *be a fixed region in* $\mathbb{R}^3$ *with boundary* S *and assume the general equations of motion hold (see* §8). *Then*

$$\frac{d}{dt}\int_W \rho \underline{v}\, dx = \int_W \rho \underline{f}\, dx + \int_S (\underline{t} - \rho(\underline{v}\cdot\underline{n})\underline{v})da . \qquad (1)$$

Proof. The equations of motion can be written

$$\frac{\partial}{\partial t}(\rho v_i) + \operatorname{div}(\rho v_i \underline{v}) = \rho f_i + T_{ij,j} ,$$

(whether or not the continuity equation holds). Integrate over W :

$$\int_W \frac{\partial}{\partial t}(\rho v_i)dx = \frac{d}{dt}\int_W \rho v_i\, dx ,$$

by the standard calculus result for fixed domain and smooth integrands;

$$\int_W \operatorname{div}(\rho v_i \underline{v})dx = \int_S \rho v_i \underline{v}\cdot\underline{n}\, da ,$$

$$\int_W T_{ij,j}\, dx = \int_S T_{ij} n_j\, da$$

$$= \int_S t_i\, da ,$$

by Gauss' theorem. Putting things together, we get the desired result. ∎

Remarks. 1. The last term on the right hand side of (1) is called the momentum flux through S .

2. Eq. (1) is a useful formula for evaluating the total force on an obstacle immersed in a fluid. For example, suppose the flow is steady (i.e., $\partial v/\partial t = 0$, $\partial \rho/\partial t = 0$) about an obstacle whose boundary is S_1 . Assume $\underline{v}$ is parallel to S_1 and let S_2 be any fixed, closed surface which contains the obstacle and some surrounding fluid (see Figure 13-1). The volume of fluid W inside S_2 has for its boundary $S = S_1 \cup S_2$. Applying (1) to this situation and further assuming that $\underline{f} = \underline{0}$, results in

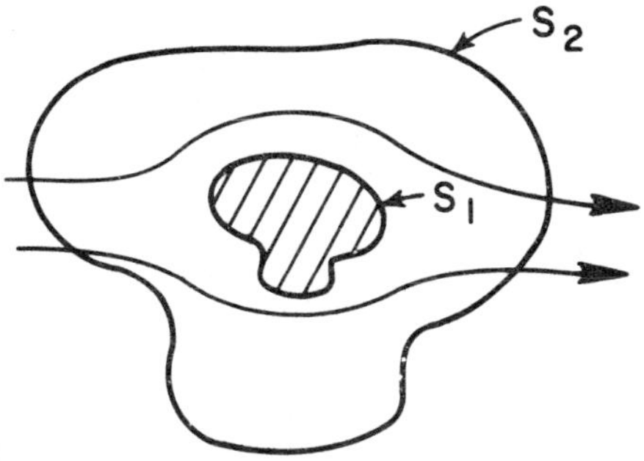

FIG. 13-1

$$\int_S (\underline{t} - \rho(\underline{v}\cdot\underline{n})\underline{v})da = \int_{S_1} \underline{t}\, da + \int_{S_2} (\underline{t} - \rho(\underline{v}\cdot\underline{n})\underline{v})da ,$$

since $\underline{v}\cdot\underline{n} = 0$ on S_1 . Therefore the total force on S_1 is given by:

$$\underline{F} = \int_{S_1} \underline{t}\, da = \int_{S_2} (\rho(\underline{v}\cdot\underline{n})\underline{v} - \underline{t})da . \qquad (2)$$

Now we are ready to discuss D'Alembert's paradox. It states the following.

In a steady, irrotational, incompressible flow of an ideal homogeneous fluid around an obstacle (occupying a "nice" bounded region) in $\mathbb{R}^3$ *having velocity* $\underline{U}$ *at* ∞ *, there is no force exerted on the obstacle.*

This result is to be compared with the two-dimensional case (§11) where lift, but no drag was possible.

The result here is essentially due to the fact that the exterior of an obstacle in 3 dimensions is simply connected. Therefore, every irrotational flow is a potential flow and hence is circulation free.

We will not present the full proof of the result hére, but we will sketch the ideas.

Recall from elementary physics that the solution of $\Delta\phi = -\rho$ in $\mathbb{R}^3$ is

$$\phi(\underline{x}) = \frac{1}{4\pi}\int \frac{\rho(\underline{y})}{\|\underline{x}-\underline{y}\|}\, dy \ ,$$

i.e., ϕ is the potential due to a charge distribution ρ . Notice that if ρ is concentrated in a finite region, then

$$\phi(\underline{x}) = 0\left(\frac{1}{r}\right) \ , \quad r = \|\underline{x}\|$$

i.e., $|\phi(\underline{x})| \leqslant \frac{\text{constant}}{r}$ as $r \to \infty$. In fact, as we know physically, $\phi(\underline{x}) \approx \frac{Q}{4\pi r}$ for r large, where $Q = \int \rho(\underline{y})dy$ is the total charge. If $Q = 0$, then $\phi(\underline{x}) = 0\left(\frac{1}{r^2}\right)$ since the first term in the expansion in powers of $\frac{1}{r}$ is now missing.

Analogously, in a flow problem, the potential ϕ satisfies

$$\Delta\phi = 0 ,$$

$$\underline{\nabla}\phi = \underline{U} \quad \text{at} \quad \infty ,$$

and

$$\frac{\partial\phi}{\partial n} = 0 \quad \text{on the boundary of the obstacle} .$$

The solution here can then be shown to satisfy

$$\phi(\underline{x}) = \underline{U}\cdot\underline{x} + O(\frac{1}{r})$$

as in the potential case above. However, we have an integral condition here, analogous to $Q = 0$. Namely, the net outflow at ∞ should be zero. This means

$$\phi(\underline{x}) = \underline{U}\cdot\underline{x} + O(\frac{1}{r^2}) .$$

Hence

$$\underline{v} = \underline{U} + O(r^{-3}) . \tag{3}$$

Since $p = \frac{-\rho}{2} v^2$, we also have $p = p_0 + O(r^{-3})$ (write $v^2 = U^2 + (\underline{v} - \underline{U})\cdot(\underline{v} + \underline{U})$ to see this).

Now according to (2), the force is

$$\underline{F} = \int_{S_2} (\rho(\underline{v}\cdot\underline{n})\underline{v} + p\underline{n})da \ .$$

We are free to choose S_2 as any sphere of radius R enclosing the obstacle. Then

$$\underline{F} = \int_{S_2} (p_0\underline{n} + \rho(\underline{U}\cdot\underline{n})\underline{U})da + (\text{area } S_2)\cdot O(R^{-3})$$

$$= 0 + O(R^{-1}) \to 0 \quad \text{as} \quad R \to \infty \ .$$

Hence $\underline{F} = \underline{0}$.

In the following problem we see that (3) is indeed valid for flow past a sphere.

<u>Problem</u>. Show that potential flow past a sphere of radius R_0 and velocity $\underline{U}$ at ∞ is given by

$$\phi = \frac{R_0^3}{2r^2}\,\underline{U}\cdot\underline{n} + \underline{x}\cdot\underline{U} \ , \quad \underline{n} = \underline{x}/\|\underline{x}\| \ ,$$

$$\underline{v} = \frac{-R_0^3}{2r^3}[\,3\underline{n}(\underline{U}\cdot\underline{n}) - \underline{U}\,] + \underline{U} \ .$$

§14. Kelvin's Circulation Theorem.

Kelvin's circulation theorem has to do with how the vorticity $\underline{\omega} = \underline{\nabla} \times \underline{v}$ is transported by the fluid. Recall that the circulation around a loop C is

$$\int_C \underline{v} \cdot \underline{ds} ,$$

and if C happens to bound a surface S in the fluid, then

$$\int_C \underline{v} \cdot \underline{ds} = \int_S \underline{\omega} \cdot \underline{dS} ,$$

by Stokes' Theorem.

Let us begin by studying the material time derivative of $\underline{\omega}$.

Theorem. *If* $\underline{v}$ *is* C^2, $D\underline{v}/Dt$ *is irrotational (i.e.,* $\underline{\nabla} \times (D\underline{v}/Dt) = \underline{0}$ *) and* $\rho, \underline{v}$ *satisfy the continuity equation, then* $\underline{\omega} = \underline{\nabla} \times \underline{v}$ *satisfies Beltrami's diffusion equation:*

$$\frac{D}{Dt}\left(\frac{\underline{\omega}}{\rho}\right) = \left(\frac{\underline{\omega}}{\rho} \cdot \underline{\nabla}\right) \underline{v}$$

Proof. Recall that

$$\frac{D}{Dt} \underline{v} = \frac{\partial \underline{v}}{\partial t} + (\underline{v} \cdot \underline{\nabla}) \underline{v}$$

and, by vector identity 7 from Appendix A,

$$(\underline{v}\cdot\underline{\nabla})\underline{v} = \underline{\omega} \times \underline{v} + \underline{\nabla}(\tfrac{1}{2} v^2) \ .$$

Taking the curl, we have

$$\underline{\nabla} \times \frac{D}{Dt}\underline{v} = \underline{\nabla} \times \frac{\partial \underline{v}}{\partial t} + \underline{\nabla} \times (\underline{\omega} \times \underline{v}) + 0 \ ,$$

$$= \frac{\partial}{\partial t}\underline{\omega} + \underline{\omega}(\underline{\nabla}\cdot\underline{v}) - \underbrace{(\underline{\nabla}\cdot\underline{\omega})\underline{v}}_{\underline{0}} + (\underline{v}\cdot\underline{\nabla})\underline{\omega} - (\underline{\omega}\cdot\underline{\nabla})\underline{v} \ ,$$

where we have used the fact that $\underline{v}$ is C^2 to interchange the order of differentiation and have employed vector identity 3 from the Appendix. Multiplying the above by ρ^{-1} and using the continuity equation enables us to compute

$$\frac{1}{\rho}(\underline{\nabla} \times \frac{D}{Dt}\underline{v}) = \frac{1}{\rho}\frac{D}{Dt}\underline{\omega} + \frac{1}{\rho}\underline{\omega}(\text{div } \underline{v}) - \frac{1}{\rho}(\underline{\omega}\cdot\underline{\nabla})\underline{v} \ ,$$

$$= \frac{D}{Dt}\left(\frac{\underline{\omega}}{\rho}\right) - \frac{1}{\rho}(\underline{\omega}\cdot\underline{\nabla})\underline{v} \ . \quad \blacksquare$$

In general $D\underline{v}/Dt$ is not irrotational. Two examples of when it is are as follows:

1. Consider an ideal fluid for which ρ is constant and $\underline{f}$ derives from a potential. We have immediately that $\underline{\nabla} \times D\underline{v}/Dt = \underline{\nabla} \times (-\frac{1}{\rho}\underline{\nabla}p + \underline{f}) \equiv \underline{0}$.

2. A (compressible) perfect fluid is <u>barotropic</u> if $p = \hat{p}(\rho)$. If this relation is invertible, then we may define a potential $h(p) = -\int^{p} d\bar{p}/\rho(\bar{p})$ such that $\underline{\nabla}h(p) = h'(p)\underline{\nabla}p = \frac{-1}{\rho}\underline{\nabla}p$. Thus, if $\underline{f}$ derives from a potential, we have again that $\underline{\nabla} \times D\underline{v}/Dt \equiv \underline{0}$.

We now draw an important consequence:

Corollary 1. *Let* $\underline{\phi}(t, \underline{x})$ *be the flow of a* C^2 *vector field* $\underline{v}$ *and assume that* ρ, $\underline{v}$ *satisfy Beltrami's diffusion equation. Then* $\underline{\omega}/\rho$ *is transported by the flow, i.e.,*

$$\frac{\underline{\omega}}{\rho}(t, \underline{\phi}(t, \underline{x})) = D\underline{\phi}_t(\underline{x})\cdot\frac{\underline{\omega}}{\rho}(0, \underline{x}) .$$

Proof. Let $\underline{F}(t, \underline{x}) = \frac{\underline{\omega}}{\rho}(t, \underline{\phi}(t, \underline{x}))$ and $\underline{G}(t, \underline{x}) = D\underline{\phi}_t(\underline{x})\cdot\frac{\underline{\omega}}{\rho}(0, \underline{x})$.

Compute

$$\frac{\partial \underline{F}}{\partial t} = \frac{D}{Dt}\left(\frac{\underline{\omega}}{\rho}\right) = \left(\frac{\underline{\omega}}{\rho}\cdot\underline{\nabla}\right)\underline{v} = \underline{F}\cdot\underline{\nabla v} ,$$

by the Beltrami diffusion equation;

$$\begin{aligned}\frac{\partial \underline{G}}{\partial t} &= D\left(\frac{\partial}{\partial t}\underline{\phi}_t(\underline{x})\right)\cdot\frac{\underline{\omega}}{\rho}(0, \underline{x}) \\ &= D(\underline{v}\circ\underline{\phi}_t)(\underline{x})\cdot\frac{\underline{\omega}}{\rho}(0, \underline{x}) \\ &= \underline{\nabla v}\cdot D\underline{\phi}_t(\underline{x})\cdot\frac{\underline{\omega}}{\rho}(0, \underline{x}) \\ &= \underline{\nabla v}\cdot\underline{G} ,\end{aligned}$$

by the properties of a C^2 flow and the chain rule. It is elucidating to express these results in coordinates:

$$\frac{\partial F_i}{\partial t} = F_j\frac{\partial v_i}{\partial x_j} , \quad \frac{\partial G_i}{\partial t} = G_j\frac{\partial v_i}{\partial x_j} .$$

Thus $\underline{F}$ and $\underline{G}$ satisfy the same linear ordinary differential equation. By hypothesis $\underline{\nabla v}$ is C^1, so from the basic existence and uniqueness theorem in §4, this equation can be uniquely solved given the initial data. Since $\underline{F}(0, \underline{x}) = \underline{G}(0, \underline{x})$ $(D\underline{\phi}_0(x) = Id)$, it follows that $\underline{F}(t, \underline{x}) = \underline{G}(t, \underline{x})$. ■

A second corollary involves the notion of a "vortex line."

Definition. A vortex line* is an integral curve of the vector field $\underline{\omega}/\rho$ for t_0 fixed, i.e., $c(s)$ is a vortex line at t_0 if $c'(s) = \frac{\underline{\omega}}{\rho}(t_0, c(s))$.

Corollary 2. *If* $\underline{\phi}(t, \underline{x})$ *is the flow in Corollary 1 and if* $c(s) = \underline{\phi}(0, c(s))$ *is a vortex line at* $t = 0$, *then* $\underline{\phi}(t, c(s))$ *is also a vortex line at time* t.

Proof.

$$\begin{aligned} \frac{\partial}{\partial s}\underline{\phi}(t, c(s)) &= D\underline{\phi}_t(c(s))\cdot c'(s) \\ &= D\underline{\phi}_t(c(s))\cdot\frac{\underline{\omega}}{\rho}(0, c(s)) \\ &= \frac{\underline{\omega}}{\rho}(t, \underline{\phi}(t, c(s))) . \end{aligned}$$

The last line follows from the previous theorem. ■

Next we ask how circulation changes in time. The following general result answers this.

---

*Sometimes, vortex line is used for a line along which the vorticity concentration is infinite. Cf. Friedrichs and Von Mises [4] p. 129.

Theorem. *Let* $\underline{v}$ *be a* C^2 *vector field and* $\underline{\phi}$ *its flow. If* C *is a closed loop of class* C^1, *then*

$$\frac{d}{dt}\int_{\underline{\phi}_t(C)} \underline{v}\cdot\underline{ds} = \int_{\underline{\phi}_t(C)} \frac{D\underline{v}}{Dt}\cdot\underline{ds}$$

Proof. Let $\underline{x}(s)$ be a parametrization of the loop C, $0 \leqslant s \leqslant 1$. A parametrization of $\underline{\phi}_t(C)$ is given by $\underline{\phi}(t, \underline{x}(s))$, $0 \leqslant s \leqslant 1$. By the change of variables formula and the chain rule, we have that:

$$\begin{aligned}
\frac{d}{dt}\int_{\underline{\phi}_t(C)} \underline{v}\cdot\underline{ds} &= \frac{d}{dt}\int_0^1 \underline{v}(t, \underline{\phi}(t, \underline{x}(s)))\cdot\frac{\partial}{\partial s}(\underline{\phi}(t, \underline{x}(s)))ds , \\
&= \frac{d}{dt}\int_0^1 \underline{v}\cdot D\underline{\phi}_t\cdot\frac{d\underline{x}}{ds}\, ds , \\
&= \int_1^0 \left(\frac{D\underline{v}}{Dt}\cdot D\underline{\phi}_t\cdot\frac{d\underline{x}}{ds} + \underline{v}\cdot D\underline{v}\cdot D\underline{\phi}_t\cdot\frac{d\underline{x}}{ds}\right)ds , \\
&= \int_{\underline{\phi}_t(C)} \left(\frac{D\underline{v}}{Dt} + \underline{v}\cdot D\underline{v}\right)\cdot\underline{ds} .
\end{aligned}$$

The second and third lines are valid since the integrand in line two is a C^1 function of t. In coordinates, the second term of the integrand, namely $\underline{v}\cdot D\underline{v}$, is $v_i\partial v_i/\partial x_j = \frac{1}{2}\underline{\nabla} v^2$. Thus, being a gradient, its integral over the C^1 loop $\underline{\phi}_t(C)$ is 0. ∎

Now we are ready to deduce:

Kelvin's Circulation Theorem. *Suppose* $\underline{f}$ *derives from a potential. Then for an ideal fluid for which* ρ = *constant, or a barotropic fluid, circulation is preserved, i.e.,* $\frac{d}{dt}\int_{\underline{\phi}_t(C)} \underline{v}\cdot\underline{ds} = 0$, *where* C *is a closed loop.*

Proof. In previous examples it was shown that, under the above hypothesis, $D\underline{v}/Dt$ was a gradient. The proof then follows by the preceding theorem. ■

Remarks. 1. Clearly the theorem holds whenever $D\underline{v}/Dt$ is a gradient in a neighborhood of $\underline{\phi}_t(C)$.

2. If C is the boundary of a simply connected smooth surface S , then by Stokes' theorem

$$\int_{\underline{\phi}_t(C)} \underline{v}\cdot\underline{ds} = \int_{\underline{\phi}_t(S)} \underline{\omega}\cdot\underline{dS} \ .$$

We emphasize that this need not be the case for the preceding theorem to hold. Kelvin's circulation theorem is valid when, for example, C goes around a hole or when C wraps around itself.

Corollary. *Suppose* $\underline{\omega} = \underline{0}$ *at* $t = 0$ *for an ideal or barotropic fluid. Then* $\underline{\omega} = \underline{0}$ *for all* t *for which the flow is defined.*

Proof. Invoke Kelvin's circulation theorem and Stokes' theorem. ■

Problem. In ideal or barotropic two-dimensional flow, show that ω/ρ , as a scalar function, is constant following the fluid (Wolibner's global existence theorem for Euler flows depends heavily upon this fact).

§15. The Helmholtz Theorems.

H. Helmholtz, in 1858 reinterpreted Kelvin's results (§14) from a more geometrical point of view. We turn now to an investigation of this aspect of the conservation of vorticity.

Definition. A vortex tube is a C^1 diffeomorphic copy* of a cylinder such that the image of each generating line of the cylinder is a vortex line (Figure 15-1). Vortex tubes are visible when the motion is circulating as shown in Figure 15-1, such as occurs in smoke rings and tornados.

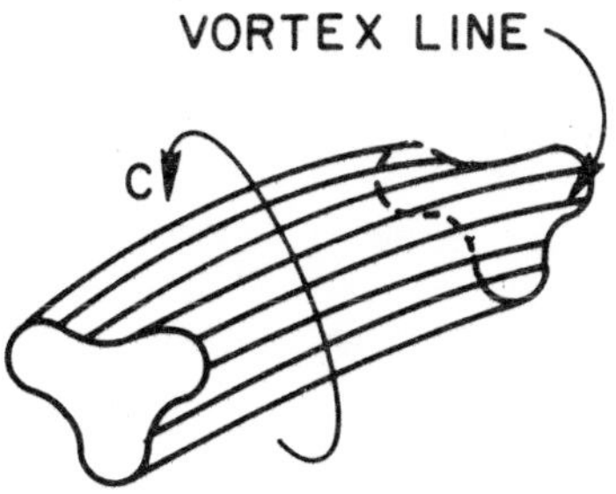

FIG. 15-1 VORTEX TUBE

Helmholtz' Theorems. *Assume the fluid in question is either ideal, with* ρ *constant, or is barotropic. Then*

(I) *If* C_1 *and* C_2 *are two curves around the vortex tube, then*

$$\int_{C_1} \underline{v}\cdot\underline{ds} = \int_{C_2} \underline{v}\cdot\underline{ds} .$$

See Figure 15-2. *Thus the strength of a vortex tube,* $\int_C \underline{v}\cdot\underline{ds}$, *where* C *is any curve around the vortex tube, is well defined.*

---

*This means that the tube is the image of a cylinder under a 1-1 C^1-mapping which has a C^1 inverse.

(II) *The strength of a vortex tube remains constant in time.*

(III) *Vortex lines are material lines, i.e., they move with the fluid.*

Proofs. (I) Let C_1 and C_2 be oriented as illustrated in Figure 15-2. The lateral surface of the vortex tube is denoted by S and the end faces are denoted by S_1 and S_2. Note that $\underline{\omega}$ is tangent to the lateral surface S. Let V denote the volume of the vortex tube and $\Sigma = S \cup S_1 \cup S_2$ denote the boundary of V. By Gauss' theorem

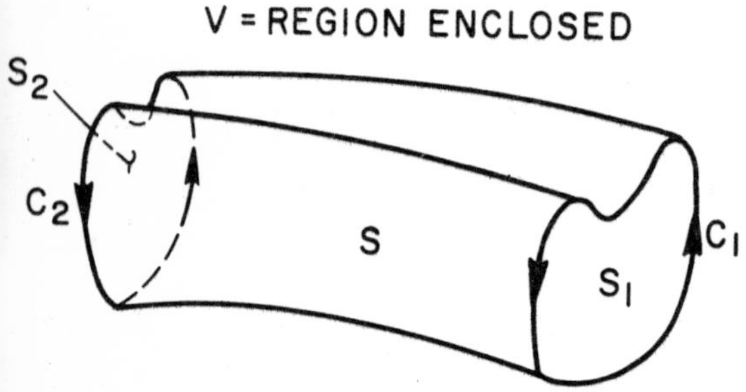

FIG. 15-2

$$0 = \int_V \underbrace{\underline{\nabla} \cdot \underline{\omega}}_{0} \, dx = \int_{\Sigma} \underline{\omega} \cdot \underline{dS} \, ,$$

$$= \int_{S_1 \cup S_2} \underline{\omega} \cdot \underline{dS} + \int_S \underbrace{\underline{\omega} \cdot \underline{dS}}_{0} \, .$$

By Stokes' theorem

$$\int_{C_1} \underline{v} \cdot \underline{ds} = \int_{S_1} \underline{\omega} \cdot \underline{dS} \, ,$$

$$\int_{C_2} \underline{v} \cdot \underline{ds} = -\int_{S_2} \underline{\omega} \cdot \underline{dS} \, ,$$

from which the result I follows.

(II) Apply Kelvin's circulation theorem.

(III) This was proved in §14. ∎

Remarks. 1. Note that the proof of Theorem I depends only upon $\nabla\cdot\underline{\omega} = 0$ in V and $\underline{\omega}$ tangent to S .

It is often stated that this result implies that a vortex line cannot end in the interior of a fluid and must, thereby form a closed loop or end on the boundary.* Actually this is only true for vortex tubes.† The intuitive reason a vortex tube cannot end in the fluid (assuming velocities do not become infinite) is seen by referring to Figure 15-3.

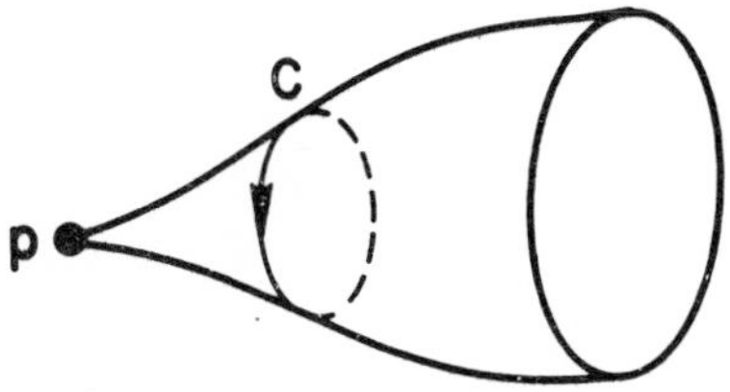

FIG. 15-3

As C moves towards p , $\int_C \underline{v}\cdot\underline{dS}$ remains constant, but C shrinks to a point, which cannot be. A compactness argument and uniqueness of

---

*H. Lamb, "Mathematical Theory of the Motion of Fluids," Cambridge (1895) p.

†O. Kellog, "Foundation of Potential Theory," Dover (1954) p. 41.

of integral curves of $\underline{\omega}$ shows that if such a tube does not end on the boundary, it must close up.

§16. Viscous Fluids and the Navier-Stokes Equations.

Thus far we have considered only perfect fluids, in which $\underline{T} = -p\underline{I}$. We now abandon this description in favor of one more in keeping with our physical notion as to how a fluid behaves, and in hope of being able to explain lift (see §§11, 13). Recall that the equations of motion for a general fluid (any continuum for that matter) are

$$\rho \frac{D\underline{v}}{Dt} = \text{div } \underline{T} + \rho \underline{f} \ . \qquad (1)$$

Thus to specify a fluid, we must give form to the way $\underline{T}$ depends on other dynamical variables, e.g., $\underline{v}$, $\underline{\nabla v}$, ρ , etc. This leads one to study constitutive theory, a branch of modern continuum mechanics. However, for the purposes of these introductory lectures, this would take us too far afield.* Thus we shall make some specific assumptions and allude to a few theorems to enable us to quickly arrive at the constitutive model which we are most interested in studying, namely the classical theory of viscous flow.

We shall assume that additional terms must be added to $-p\underline{I}$ as follows:

$$\underline{T}(t, \underline{x}) = -p(t, \underline{x})\underline{I} + \underline{\tilde{T}}(\underline{D}(t, \underline{x})) \qquad (2)$$

---

*See, e.g. C. Truesdell, "The Elements of Continuum Mechanics," Springer-Verlag (1966).

where

1. $\tilde{\underline{T}}$ is a smooth function of $\underline{D}$.

2. $\tilde{\underline{T}}$ is <u>isotropic</u>, i.e., for all orthogonal matrices $\underline{U}$

$$\tilde{\underline{T}}(\underline{U}\,\underline{D}\,\underline{U}^{-1}) = \underline{U}\,\tilde{\underline{T}}\,(\underline{D})\,\underline{U}^{-1} .$$

and

3. $\tilde{\underline{T}}(\underline{0}) = \underline{0}$.

Heuristically, we think of $\tilde{\underline{T}}$ as the stresses due to viscous phenomena. We remark that basic hypotheses of constitutive theory imply that $\tilde{\underline{T}} = \tilde{\underline{T}}(\underline{D})$ even if we had started with the more general assumption that $\tilde{\underline{T}} = \tilde{\underline{T}}(\underline{v}, \underline{\nabla v})$.

The intuitive reason that $\tilde{\underline{T}}$ depends only on $\underline{D}$ is this: recall from §12 that changes in $\underline{v}$ can be decomposed into rotations $\underline{\Omega}$ (the skew part of $\underline{\nabla v}$) and deformations $\underline{D}$ (the symmetric part of $\underline{\nabla v}$) . Viscous forces are present only if velocity gradients are present and should not depend on local rotations of the fluid. Thus they should depend only on the deformation part, $\underline{D}$.

<u>Theorem</u>. *Assume* 1., 2. *and* 3. *hold. Then*

$$\tilde{\underline{T}} = \alpha\underline{I} + \beta\underline{D} + \gamma\underline{D}^2 \tag{3}$$

where α, β *and* γ *are scalar functions of the invariant of* $\underline{D}$ *and* $\alpha = 0$ *if* $\underline{D} = \underline{0}$.

(The invariants of $\underline{D}$ are the coefficients in its characteristic polynomial.)

A proof is contained in Serrin [11].

One can give a plausibility argument as follows: If higher-order terms like $\underline{D}^3$ occur, we could reexpress them in terms of $\underline{I}$, $\underline{D}$ and $\underline{D}^2$ since $\underline{D}$ satisfies its characteristic equation (Cayley-Hamilton theorem). The coefficients depend only on the invariants since these are the only rotationally invariant scalar functions of $\underline{D}$. Thus the fact that we have $\underline{D}^2$ in (3) depends on the fact that we are in dimension 3.

In most of classical fluid mechanics it is assumed that $\underline{\tilde{T}}$ is linear in $\underline{D}$, i.e.,* $\gamma \equiv 0$, β is constant and α is linear in $\underline{D}$.

We note that the interpretation of (2) is distinctly different for compressible and incompressible fluids, just as in the case for perfect fluids. In the compressible case, p is related to other dynamical variables, e.g., ρ . In the incompressible case, p is a dynamical variable to be solved for by the equations of motion and $\operatorname{div} \underline{v} = 0$.

If we assume $\underline{\tilde{T}}$ is linear in $\underline{D}$, then $\alpha = \lambda(\operatorname{trace} \underline{D})$ for a constant λ , since this is the only linear scalar function of the invariants of $\underline{D}$. Thus, calling $\beta = 2\mu$, we have

$$\underline{\tilde{T}} = \lambda\ (\operatorname{div} \underline{v})\ \underline{I} + 2\mu\ \underline{D}\ ,$$

where λ, μ are constants (note that $\operatorname{div} \underline{v} = \operatorname{trace} \underline{D}$) . μ is known as the <u>dynamical</u> or <u>shear viscosity</u>. When $\underline{T} = -p\underline{I} + \underline{\tilde{T}}$ is substituted into the equations of motion, we arrive at the <u>Navier-Stokes equations</u>:

---

*The importance of γ does not seem to be well understood. Most practicing fluid dynamicists ignore γ . For most fluids it is very small, but some have argued that it is important in high velocity flows.

(compressible)

$$\rho \frac{D\underline{v}}{Dt} = \rho\underline{f} - \underline{\nabla}p + (\lambda + \mu)\,\underline{\nabla}\,(\mathrm{div}\,\underline{v}) + \mu\,\Delta\,\underline{v}\ , \qquad (4)$$

(incompressible)

$$\frac{D\underline{v}}{Dt} = \underline{f} - \frac{1}{\rho}\,\underline{\nabla}p + \nu\,\Delta\,\underline{v}\ , \qquad (4')$$

where $\nu = \mu/\rho$ is the <u>kinematic</u> <u>viscosity</u>. When ρ is constant in the incompressible case we can incorporate ρ as part of p , since p is unknown anyway, the result being that ρ is then absent in the equations (4′).

In the sequel we shall deal only with the so-called no-slip boundary condition, in which $\underline{v}$ equals the prescribed velocity of the boundary. For stationary boundaries this condition becomes $\underline{v} = \underline{0}$. This is to be contrasted with the condition we assumed for a perfect fluid, i.e., that $\underline{v}\cdot\underline{n} = 0$ on a stationary boundary. The mathematical reasons why these boundary conditions are appropriate for their respective equations involve consideration of the well-posedness of initial-boundary value problems and are a delicate matter.

For second-order equations like (4) or (4′) one needs the same number of boundary conditions as equations, in this case three; one for each component of $\underline{v}$. Experimental work indicates the veracity of the no-slip condition. (See Birkhoff [1], Shinbrot [13] and Schnute and Shinbrot, <u>Kinetic Theory and Boundary Conditions for Fluids</u>, Can. J. Math. <u>25</u> (1973) 1183-1215 for further discussions.)

Incompressible Viscous Fluids. For awhile we shall be concerned with (4′) when ρ = constant.

Recall that the stress on a material surface is given by $T_{ij}n_j$, which for the present case becomes $T_{ij}n_j = -pn_i + \tilde{T}_{ij}n_j = -pn_i + 2\mu D_{ij}n_j$.

Example. Let the boundary of the fluid be given by the upper half-plane (Figure 16-1). We compute

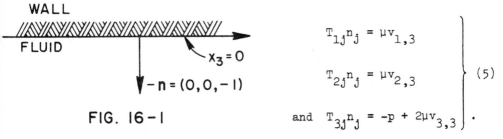

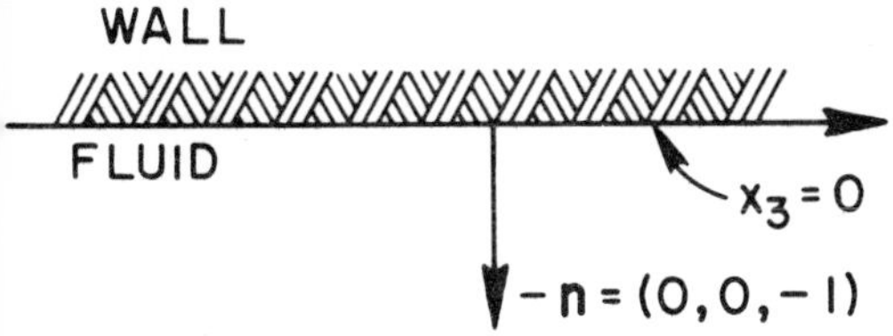

FIG. 16-1

$$\left.\begin{aligned} T_{1j}n_j &= \mu v_{1,3} \\ T_{2j}n_j &= \mu v_{2,3} \\ \text{and}\quad T_{3j}n_j &= -p + 2\mu v_{3,3} \end{aligned}\right\} \quad (5)$$

(5) are the components of the traction vector "with respect to the fluid" (recall $\underline{n}$ is the outward normal vector). The components of the traction vector "with respect to the wall" are the negatives of (5). We sometimes refer to these as the tractions on the wall.

Note that when $\underline{v} = \underline{0}$ on a smooth surface $\underline{\omega}\cdot\underline{n} = 0$. This can be easily seen in local coordinates, viz., let x_3 represent the normal direction to the surface in question and let x_1, x_2 be the tangential directions. Clearly $v_{1,2} - v_{2,1} \equiv 0$. At each point on the surface a coordiante system can be constructed such that $\omega_3 = v_{1,2} - v_{2,1}$, thus verifying the assertion.

Definition. A stationary incompressible viscous flow consists of a pair $\underline{v}$ and p, independent of t, such that

$$\left.\begin{aligned} \rho(\underline{v}\cdot\underline{\nabla})\underline{v} &= -\underline{\nabla}p + \mu\,\Delta\,\underline{v} \\ \operatorname{div}\,\underline{v} &= 0\ . \end{aligned}\right\} \qquad (6)$$

We now consider some cases where (6) can be solved exactly.

Example 1. Flow between two infinite plates.

The geometrical setup is indicated in Figure 16-2. The top plate has constant velocity $(u, 0, 0)$ and the bottom one is fixed. Thus the no-slip boundary conditions are

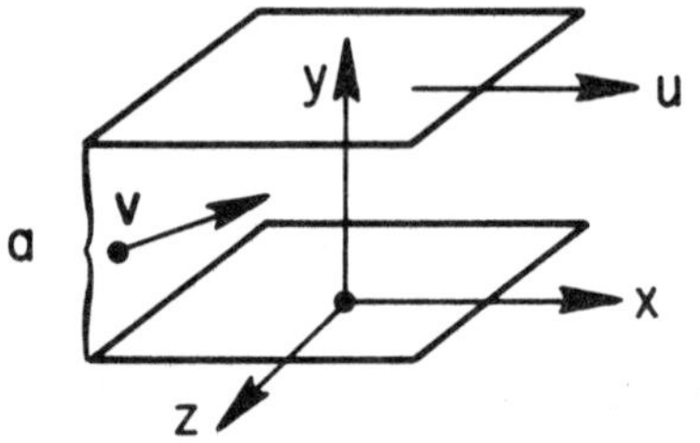

FIG. 16-2

$$\underline{v} = (u, 0, 0)\ , \quad y = a\ ,$$

$$\underline{v} = \underline{0}\ , \quad y = 0\ .$$

Assume the solution takes the form $v_x = v_x(y)$, $v_y = v_z = 0$ and $p = p(y)$. Substituting these into (6), we see that the z Navier-Stokes equation and the incompressibility condition are satisfied identically. The x equation becomes

$$\frac{\partial^2 v_x}{\partial y^2} = 0\ ,$$

which integrates to $v_x = Ay + B$, where A, B are constants. Imposition of the boundary conditions yields $v_x = (u/a)y$ (Figure 16-3). The y Navier-Stokes equation tells us that $p = \text{constant}$. Note that if we had been working with the Euler equations instead of the

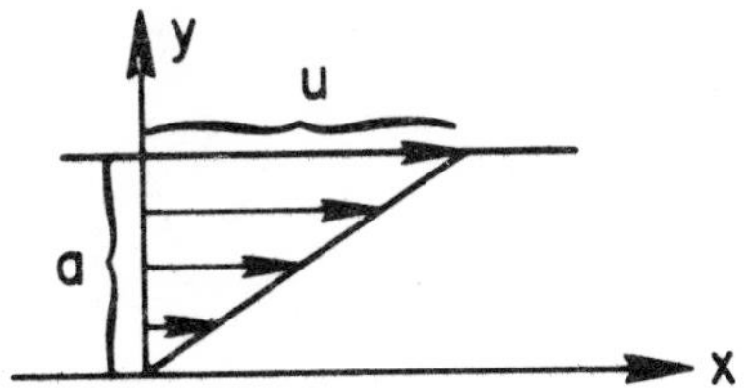

FIG. 16-3

Navier-Stokes equations, <u>any</u> $v_x(y)$, $v_y = v_z = 0$ and p = constant, would constitute a solution.

<u>Problem</u>. Show that the normal traction with respect to the top plate is just p and that the x-tangential traction with respect to the top plate is $-\mu u/a$.

<u>Example 2</u>. Flow between two stationary plates in the presence of a pressure gradient.

The geometry is the same as in Figure 16-2, but here the top plate is also stationary. Thus the no-slip boundary conditions are $\underline{v} = \underline{0}$ for $y = 0$ and a . Assume the solution takes the form $v_x = v_x(y)$, $v_y = v_z = 0$ and $p = Cx$, where C is a constant. The solution satisfies the y and z Navier-Stokes equations and the incompressibility condition identically. The x Navier-Stokes equation becomes

$$C = \mu \frac{\partial^2 v_x}{\partial y^2} ,$$

which integrates to $v_x = (C/2\mu)y^2 + Ay + B$, A, B constants. Taking account of the boundary conditions, we have

$$v_x = -\frac{C}{2\mu}\left\{ \frac{a^2}{4} - \left(y - \frac{a}{2}\right)^2 \right\} ,$$

which is depicted in Figure 16-4.

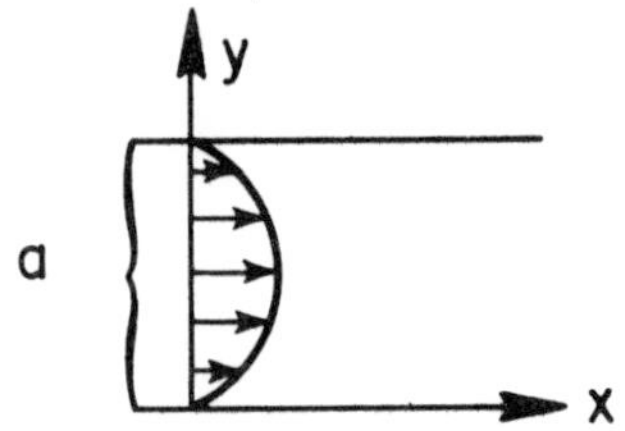

FIG. 16-4

<u>Problem</u>. Compute the tractions on the plates.

<u>Example 3</u>. Flow in an infinite pipe.

The geometry is illustrated in Figure 16-5. We work in cylindrical coordinates. The no-slip boundary condition is that $\underline{v} = \underline{0}$ when $r = a$. Assume the solution takes the form

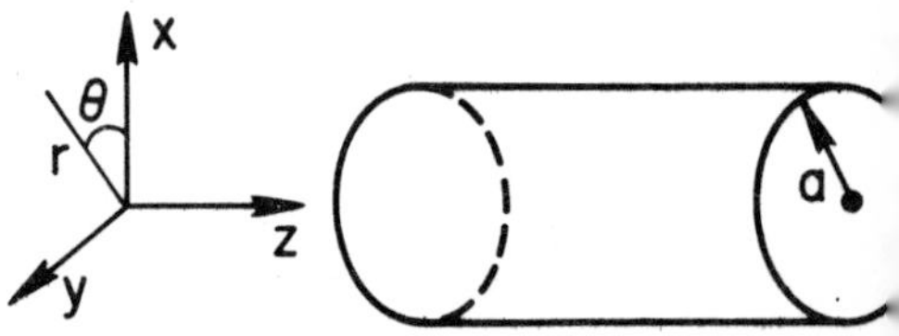

FIG. 16-5

$p = Cz$, C constant, $v_z = v_z(r)$

and $v_r = v_\theta = 0$. Substituting these into the cylindrical coordinate form of (5) (see the Appendix), we get that the r and θ Navier-Stokes equations and the incompressibility condition are identically satisfied. The z Navier-Stokes equation becomes

$$C = \mu \, \Delta v_z = \mu\left(\frac{1}{r}\frac{\partial}{\partial r}\left(r \frac{\partial v_z}{\partial r}\right)\right) .$$

Integration yields $v = -(C/4\mu)r^2 + A \log r + B$, where A, B are constants. Since we require that the solution is bounded, A must be 0 , since $\log r \to \infty$ as $r \to 0$. The no-slip condition determines B :

$$v_z = \frac{C}{4\mu}(a^2 - r^2) .$$

The solution is depicted in Figure 16-6. This example is the famous Poiseuille (pwa zwē) flow. It is of interest to compute the mass flow rate $Q = \int_S \rho v_z \, da$ through the pipe , viz.,

$$\begin{aligned} Q &= \int_S \rho v_z \, da \\ &= 2\pi\rho \int_0^a v_z \, r dr \\ &= \frac{\pi C}{8\nu} a^4 . \end{aligned}$$

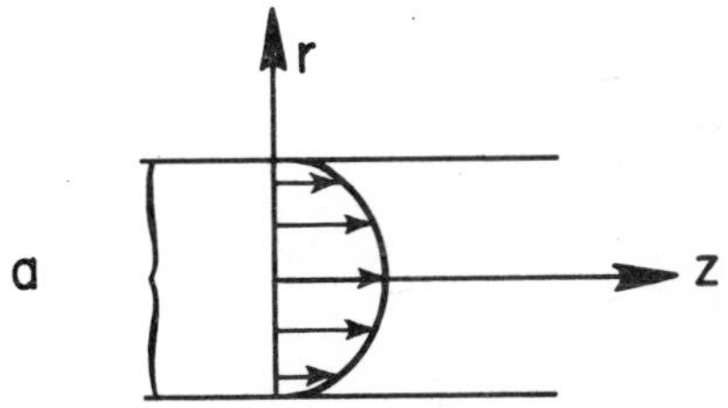

FIG. 16-6

This is the so-called fourth-power law.

Problem. Compute the tractions on the walls.

Problem. Compute the solution to the problem of stationary viscous flow between two concentric cylinders. Determine the tractions on the walls. Hint: proceed as above but retain the log term.

Another flow that can be solved explicitly is Couette flow (see

§12) , which solves coincidentally both the Euler and Navier-Stokes equations. Finally we mention that one can also solve for the flow in a converging or diverging channel (a modification of Examples 1 and 2). This is harder and was done by Hamel in 1916. (See Landau-Lifschitz [5] or Shinbrot [12] for presentations).

Addendum. Passage to the Incompressible Limit in Terms of a Material Parameter.

Here we show how to obtain the constitutive equation for an incompressible fluid as the limit of that for a compressible one. The procedure emphasizes the vastly different mathematical nature of the pressure in compressible and incompressible fluid dynamics. Our starting point is the constitutive relation for a viscous compressible fluid from this section:

$$\underline{T} = -p\underline{I} + \lambda(\mathrm{tr}\ \underline{D})\underline{I} + 2\mu\underline{D} , \tag{7}$$

where $p = \hat{p}(\rho)$. Let $f: \mathbb{R}^6 \to \mathbb{R}$ be defined by

$$f(\underline{D}) = -p\ \mathrm{tr}\ \underline{D} + \frac{\lambda}{2}(\mathrm{tr}\ \underline{D})^2 + \mu\ \mathrm{tr}\ \underline{D}^2 .$$

Then

$$T_{ij} = \partial f/\partial D_{ij} .$$

Thus the stress components T_{ij} may be thought of as the variables "conjugate" to D_{ij} with respect to the potential f. Let $\hat{f}(\hat{\underline{D}}, \text{tr } \underline{D}) = f(\underline{D})$, where $\hat{\underline{D}} = \underline{D} - \frac{1}{3}(\text{tr } \underline{D})\underline{I}$ is the deviatoric part of the deformation tensor. Note that $\underline{D} \mapsto (\hat{\underline{D}}, \text{tr } \underline{D})$ is an isomorphism. Define a pressure-like quantity $h = -D_2\hat{f}$, the derivative with respect to $\text{tr } D$. Then (7) may be replaced by the pair of equations

$$\left.\begin{aligned} -h &= -p + (\lambda + \tfrac{2}{3}\mu)\text{tr } \underline{D} \\ \text{and} \qquad \underline{T} &= -h\underline{I} + 2\mu\hat{\underline{D}} . \end{aligned}\right\} \qquad (8)$$

The term $\lambda + \frac{2}{3}\mu$ is sometimes called the bulk modulus in continuum mechanics. As long as $(\lambda + \frac{2}{3}\mu) \neq 0$, $(8)_1$ may be solved for $\text{tr } \underline{D}$ in terms of h:

$$\text{tr } \underline{D} = (-h + p)/(\lambda + \tfrac{2}{3}\mu) . \qquad (9)$$

In this way we may view $(\hat{\underline{D}}, h)$ as independent variables instead of $(\hat{\underline{D}}, \text{tr } \underline{D})$.

The incompressible limit is characterized by $\rho \to \rho_0$ and $\text{tr } D \to 0$; see §9. We assume that $\hat{p}$ is fixed throught and that $\hat{p}(\rho_0) < \infty$. To ensure $\text{tr } \underline{D} \to 0$, (9) requires that $(\lambda + \frac{2}{3}\mu) \to \pm\infty$. In addition, we require that $(8)_2$ make sense in this limit, i.e., that μ remain bounded. Thus it must be that $\lambda \to \pm\infty$ and $(8)_2$ becomes the constitutive equation for the incompressible case.

In summary: let $\hat{p}$ be given; fix the dynamic viscosity μ and solve the compressible fluid equations using T given by (8) $(p = \hat{p}(\rho))$. Then take the limit $\lambda \to \infty$. In this limit, h should converge to the pressure for the limiting incompressible flow and $\text{tr}\,\underline{D} \to 0$.

The standard way of writing the constitutive equation for an incompressible fluid is

$$\underline{T} = -p\underline{I} + 2\mu\underline{D} \ , \tag{10}$$

where p is an unknown function, i.e., h in $(8)_2$. One then require separately that $\text{tr}\,\underline{D} = 0$. This is equivalent to requiring $\underline{D} = \hat{\underline{D}}$. Using the same notation p in both (7) and (10) is a customary, but unfortunate, practice as can be deduced from the arguments above. One should observe that $\frac{1}{3}\,\text{tr}\,\underline{T}$, the mean stress, is equal to $-h$ (cf. $(8)_1$), not $-\hat{p}$, in the compressible case. This can and does lead to considerable confusion in experimental work.

The above procedure explains the correct way to make sense out of the incompressible limit. One would like to prove that the solutions of the equations converge in this limit. In this direction, results are scarce; the only rigorous theorems we know of are due to D. Ebin Motion of a Slightly Compressible Fluid, Proc. Nat. Acad. Sci. 72 (197 539-542, J.L. Lions and R. Temam (see C.R. Acad. Sci. Paris, 262 (1966 219-221). For the numerical end of this technique, see A. Chorin, Numerical Solution of Incompressible Flow Problems, Studies in Numeric Analysis 2 (1968) 64-71.

§17. The Reynolds Number Re .

Let U be a typical velocity in a flow (such as mainstream velocity or the velocity of the boundary, etc.) and let L be a typical length (such as the diameter of a region or body, etc.). Then the dimensionless number $Re = UL/\nu$ is called the Reynolds number.

U and L determine a typical time $T = L/U$. Let $\bar{U}$, $\bar{L}$, $\bar{\nu}$ be another set of typical numbers and define

$$\underline{\bar{x}} = \frac{\bar{L}}{L}\,\underline{x}\,, \quad \bar{t} = \frac{\bar{T}}{T}\,t \quad \text{and} \quad \underline{\bar{v}}(t, \underline{\bar{x}}) = \frac{\bar{U}}{U}\,\underline{v}(t, \underline{x})\,.$$

With these we have the following result:

Theorem (Reynolds 1883). *Let* $\underline{v}$ *satisfy*

$$\begin{aligned} \frac{\partial \underline{v}}{\partial t} + (\underline{v}\cdot\underline{\nabla})\underline{v} &= -\,\underline{\nabla}\,p + \nu\Delta\,\underline{v}\,, \\ \underline{\nabla}\cdot\underline{v} &= 0\,, \end{aligned} \tag{1}$$

for some function p .

Then if $\bar{Re} = Re$, $\underline{\bar{v}}$ *satisfies*

$$\begin{aligned} \frac{\partial \underline{\bar{v}}}{\partial \bar{t}} + (\underline{\bar{v}}\cdot\underline{\bar{\nabla}})\underline{\bar{v}} &= -\,\underline{\bar{\nabla}}\,\bar{p} + \bar{\nu}\,\bar{\Delta}\,\underline{\bar{v}}\,, \\ \underline{\bar{\nabla}}\cdot\underline{\bar{v}} &= 0\,, \end{aligned} \tag{2}$$

where $\bar{p} = \left(\frac{\bar{U}\bar{L}}{U\bar{L}}\right)^2 p$. *In this case we say* $\underline{v}$ *and* $\underline{\bar{v}}$ *are similar.*

Proof. The proof is just a straightforward verification (do it!). ■

Remarks. 1. This result is sometimes known as the law of similarity. It is useful for scaling results, e.g., if we wanted to experimentally model a flow governed by (1) we could build a model in a laboratory governed by (2) with $\bar{Re} = Re$ and relate the results to those of (1) via the scale factors.

2. Similar flows can also be constructed in the presence of body forces. For example, if gravitational forces are included the Froude number $F = U^2/Lg$, where g is gravitational acceleration, becomes important. The flows $\underline{v}$ and $\bar{\underline{v}}$ are similar in this case if $(Re, F) = (\bar{Re}, \bar{F})$. Another important example applies to periodic (τ = period) body forces or boundary data. Here, similarity is achieved if $(Re, S) = (\bar{Re}, \bar{S})$, where $S = U\tau/L$ is the Strouhal number.

The limiting cases, $Re \to 0$ and $Re \to \infty$, are both interesting. In a sense, the former is "easy" whereas the latter is "hard". This will be developed as we proceed. For the time being we consider the "easy" case $Re \to 0$, in which (1) becomes Stokes' equations:

$$\begin{aligned} \nu \Delta \underline{v} &= \operatorname{grad} p , \\ \operatorname{div} \underline{v} &= 0 , \end{aligned} \qquad (3)$$

and $\underline{v}$ is to satisfy a no-slip boundary condition. Under reasonable conditions there exists a unique solution to (3)*. We consider a particular case.

Let a spherical object move through a fluid in $\mathbb{R}^3$. For very slow velocities we shall assume Stokes' equations apply. We take the point of view that the object is stationary and the fluid streams by. The setup for the boundary value problem is, given $\underline{U} = (U, 0, 0)$, U constant, find $\underline{v}$ and p such that (3) hold in the region exterior to a sphere of radius R, $\underline{v} = 0$ on the boundary of the sphere and $\underline{v} = \underline{U}$ at infinity. The solution to this problem (in spherical coordinates centered in the object) is called Stokes' Flow:

$$\underline{v} = -\frac{3}{4} R \frac{\underline{U} + \underline{n}(\underline{U}\cdot\underline{n})}{r}$$

$$-\frac{1}{4} R^3 \frac{\underline{U} - 3\underline{n}(\underline{U}\cdot\underline{n})}{r^3} + \underline{U} ,$$

$$p = p_0 - \frac{3}{2} \nu \frac{\underline{U}\cdot\underline{n}}{r^2} R ,$$

where p_0 is an arbitrary constant and $\underline{n} = \frac{\underline{r}}{r}$. The student should verify this solution. A discussion of how it is obtained is contained in Landau-Lifshitz [5] and Shinbrot [13]. A picture of the stream lines (= particle paths here) is given in Figure 17-1.

---

*See Ladyzhenshaya [12] and Finn, Mathematical Questions Relating to Viscous Fluid Flow in an Exterior Domain, Rocky Mountain J. Math. 3(1973) 107-140. This last article is a good reference for what is known to date about existence and uniqueness for stationary flows.

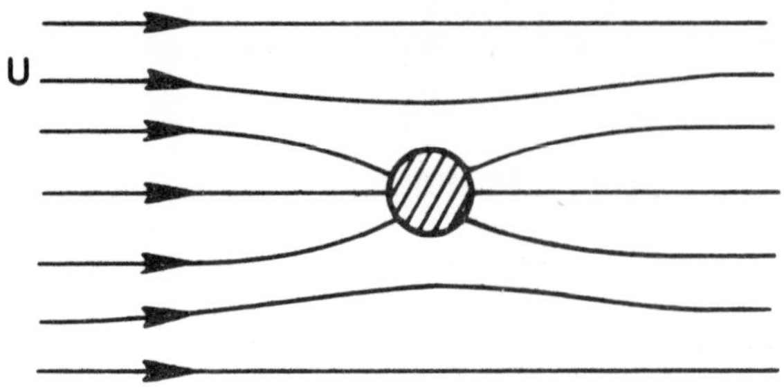

FIG. 17-1

The drag on the sphere is $F = 6\pi R\mu U$ and there is no lift (verify this by direct computation).

Physically speaking Figure 17-1 looks a little strange as it is symmetric about the vertical plane. Other amusing aspects of this solution are that

1. $\|(\underline{v}\cdot\underline{\nabla})\underline{v}\|/\|\nu\Delta\underline{v}\| \to \infty$ as $\underline{x} \to \infty$, thus neglecting the $(\underline{v}\cdot\underline{\nabla})\cdot\underline{v}$ term in the Navier-Stokes equations cannot be justified a posteriori.

2. There is a net outflow at infinity, i.e., there is an infinite wake. This is due to the presence of the $1/r$ term. Compare potential flow in §10.

Oseen (1910) suggested that Stokes' equations be replaced by

$$-\nu\Delta\underline{u} + (\underline{U}\cdot\underline{\nabla})\underline{u} = -\frac{1}{\rho}\operatorname{grad} p \ ,$$

$$\operatorname{div}\underline{u} = 0 \ ,$$

where $\underline{u} = \underline{v} - \underline{U}$. This amounts to linearizing the Navier-Stokes equations about $\underline{U}$ whereas Stokes' equations may be viewed as a linearization about $\underline{0}$. One would thus conjecture that Oseen's equations are good where the flow is close to the free stream velocity $\underline{U}$ (away from the sphere) and that Stokes' equations are good where the velocity is $\underline{0}$ (near the sphere). This is indeed the case. The solution of Oseen's equations in the region exterior to a sphere in $\mathbb{R}^3$ can be found in Lamb [7]. The drag on the sphere for the Oseen solution is $F = 6\pi RU\mu(1 + \frac{3}{8}\,\mathrm{Re})$, where $\mathrm{Re} = UR/\nu$ is the Reynolds number. Thus there is a difference of the order Re in the Stokes and Oseen drag forces.

Remarks. 1. If Ω is bounded with smooth boundary, then there exists a unique solution to Stokes' equations. (Ladyzhenskaya [12]).

2. On the exterior of a bounded region in $\mathbb{R}^3$ there exists a unique solution to Stokes' equations. The situation in $\mathbb{R}^2$ is different. In fact, we have the following strange situation:

Stokes' Paradox. There is no solution to Stokes' equations in $\mathbb{R}^2$ in the region exterior to a disk (with decent boundary conditions).*

Stokes' paradox does not apply to the Oseen or Navier-Stokes equations in $\mathbb{R}^2$ or $\mathbb{R}^3$. However, we have

*See Birkhoff [1], and J. Heywood, Arch. Rat. Mech. 37(1970) 48-60, Acta Math. 129(1972) 11-34.

Filon's Paradox (1927). Filon says that Oseen's equations are no good too. The example he gives is a skewed ellipse in a free stream (Figure 17-2). Computation of the moment exerted on the ellipse reveals that

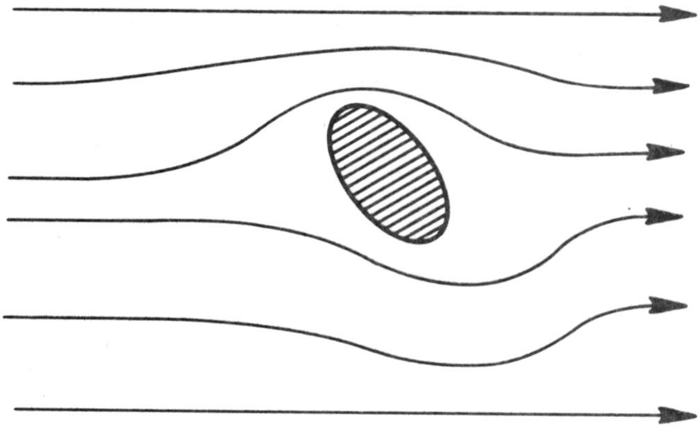

FIG. 17-2

it is infinite. This is not so surprising in view of the fact that Oseen's equations represent linearization about the free stream. One cannot expect them to give good results around the obstacle since the equations contain errors there of order U^2 .

Remarks on Existence and Uniqueness. The best single reference is Ladyzhenskaya [12] for both the stationary and non-stationary case. (See also Shinbrot [13]). For the stationary case, see the preceding footnote. For the full Navier-Stokes equations, there exists (in a certain technical function space) a unique solution with given initial data v_0 and boundary conditions, at least on a short time interval $[0, T]$. The T gets smaller and smaller as $\underline{v}_0$ gets bigger or as Re gets bigger. Specifically, the existing theorems do not decide whether or not the equations, or rather their solutions, "break down" when something interesting like turbulence is present. Further comments on recent work on this problem are offered in Section 23.

§18 Stability, Bifurcation and Large Reynolds Numbers.

In the last section we discussed the limit $Re \to 0$; i.e., slow or very viscous flow. Now we turn to the other extreme, the limit $Re \to \infty$, i.e., very fast or only slightly viscous flow.

The situation for large Re is more complicated and is the subject of much research by mathematicians and engineers alike. Here we shall give the basics of the subject, returning to it again in a brief discussion of turbulence in §22.

The concept of stability plays a central role in this discussion so we begin with this concept.

Let P be $\mathbb{R}^n$ (or more generally a Banach or Hilbert space consisting of functions whose domain and range is $\mathbb{R}^n$) and let $\underline{F}_t$ denote a (mathematical) flow on P (cf. §4). Let $\underline{X}$, a vector field on P, be the *infinitesimal generator* of $\underline{F}_t$, i.e.,

$$\frac{d}{dt}\underline{F}_t(\underline{x}) = \underline{X}(F_t(\underline{x}), t) \ , \quad \underline{x} \in P \ . \tag{1}$$

For our present purposes, we assume $\underline{X}$ is given and thus (1) constitutes a system of differential equations which we may solve for $\underline{F}_t$. In this regard we assume that we have a good existence and uniqueness theorem for (1). If $\underline{X}$ is independent of t we say it is *autonomous* and if $\underline{x}_0$ is such that $\underline{X}(x_0) = 0$ we call $\underline{x}_0$ a *fixed point* of $\underline{X}$. It follows immediately from (1) and the initial condition $\underline{F}_0(\underline{x}) = \underline{x}$ for all $\underline{x} \in P$, that if $\underline{x}_0$ is a fixed point of $\underline{X}$, then

$\underline{F}_t(\underline{x}_0) = \underline{x}_0$.

Definition. $\underline{x}_0$ is an asymptotically stable fixed point of the vector field $\underline{X}$ if there exists a neighborhood $U \subset P$ of $\underline{x}_0$ such that if $\underline{x} \in U$ then $\underline{F}_t(\underline{x}) \to \underline{x}_0$ as $t \to \infty$. In the sequel we shall refer to such a point simply as a stable point. If one can choose $U = P$, we say that $\underline{x}_0$ is globally stable.

Liapunov Stability Theorem.* *Let* $\underline{x}_0$ *be a fixed point of a smooth autonomous vector field* $\underline{X}$ *on* $P = \mathbb{R}^n$. *Then* $\underline{x}_0$ *is a stable point if the eigenvalues of the* $n \times n$ *matrix* $D\underline{X}(\underline{x}_0)$ *of partial derivatives of* $\underline{X}$ *have real parts* < 0 .

Example. Suppose we have a linear autonomous system, i.e., $\underline{X} = \underline{A}$, an $n \times n$ real matrix, which we assume is diagonalizable. Then the solution of

$$\frac{d}{dt}\underline{F}_t(\underline{x}) = \underline{A}\cdot\underline{F}_t(\underline{x}) \ , \quad \underline{F}_0(\underline{x}) = \underline{x} \ ,$$

is $\underline{F}_t(\underline{x}) = e^{t\underline{A}}\underline{x}$ where $e^{t\underline{A}} = \sum_{i=0}^{\infty} t^i\underline{A}^i/i!$. This series is absolutely convergent. Transforming to canonical coordinates:

*An excellent reference for this topic and related subjects is the recent book by M. Hirsch and S. Smale: "Differential Equations, Dynamical Systems, and Linear Algebra," Academic Press (1974).

$$\underline{A} \longrightarrow \begin{pmatrix} \lambda_1 & & & \text{zeroes} \\ & \lambda_2 & & \\ & & \ddots & \\ \text{zeroes} & & & \lambda_n \end{pmatrix},$$

and

$$e^{t\underline{A}} \longrightarrow \begin{pmatrix} e^{t\lambda_1} & & & \text{zeroes} \\ & e^{t\lambda_2} & & \\ & & \ddots & \\ \text{zeroes} & & & e^{t\lambda_n} \end{pmatrix},$$

where $\lambda_1, \ldots, \lambda_n$ are the eigenvalues of $\underline{A}$ (they occur in conjugate p[airs] since $\underline{A}$ is real). This indicates why the stability theorem is true, viz., if $\mathrm{Re}(\lambda_i) < 0$, $i = 1, \ldots, n$, then the latter matrix above converges to the zero matrix as $t \to \infty$. Observe that for the linear case consideration of neighborhoods is unnecessary.

The preceding example is helpful in understanding the nonlinear case as well. To see this expand $\underline{X}$ by Taylor's formula about $\underline{x}_0$:

$$\underline{X}(\underline{x}) = \underbrace{\underline{X}(\underline{x}_0)}_{0} + D\underline{X}(\underline{x}_0)\cdot(\underline{x} - \underline{x}_0) + \underline{\theta}(\underline{x} - \underline{x}_0) ,$$

where $\underline{\theta}(\underline{x} - \underline{x}_0)$ is $o(\|\underline{x} - \underline{x}_0\|)$ and $\|\cdot\|$ is the norm on P. Thus for $\underline{x}$ sufficiently close to $\underline{x}_0$ the term $D\underline{X}(\underline{x}_0)$ dominates the behavior of the flow of $\underline{X}$.

Remark. The stability theorem requires that the spectrum of $D\underline{X}(\underline{x}_0)$ lie entirely in the left half-plane of the complex plane $\mathbb{C}$ for $\underline{x}_0$

to be stable (see Figure 18-1).

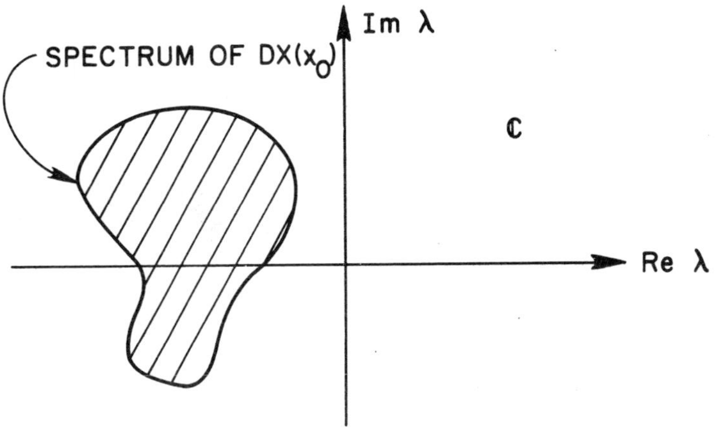

FIG. 18-1

Problem. Consider the vector field

$$\underline{X}_{\mu}(x, y) = (y, \mu(1 - x^2)y - x)$$

on $\mathbb{R}^2$. Note that (0, 0) is a fixed point. Determine if (0, 0) is stable or unstable for various values of μ.

We shall now examine how the above concepts enable us to study the stability of solutions of the Navier-Stokes equations. Let's consider the example of flow in a pipe and assume $\underline{v}_0$ is a stationary solution of the Navier-Stokes equations, e.g., the Poiseuille solution (Fig. 18-2). We are interested in what happens when $\underline{v}_0$ is perturbed, i.e., when $\underline{v}_0 \to \underline{v}_0 + \underline{\delta v}$. Note that the disturbed solution of the Navier-Stokes equations, $\underline{v}_0 + \underline{\delta v}$, will not be a stationary solution. This is part

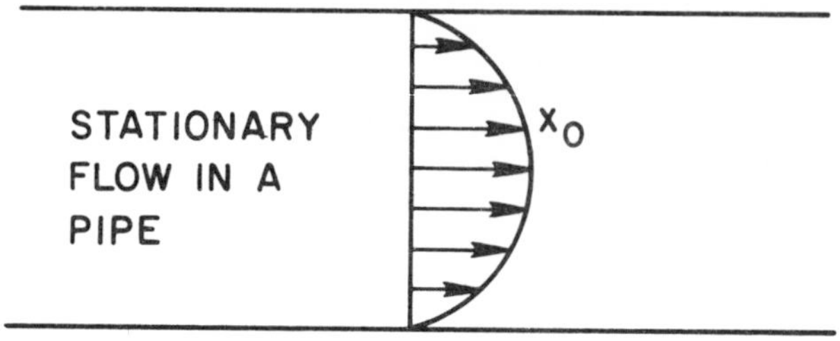

FIG. 18-2

of our notion of stability, i.e., $\underline{v}_0$ is stationary corresponds to $\underline{v}_0$ being a fixed point of a vector field on the space of all possible solutions of the Navier-Stokes equations which we shall denote by P . For the present example P would consist of divergence-free velocity fields satisfying the appropriate boundary conditions. The Navier-Stokes equations determine the dynamics on this space P of velocity fields. We write

$$\frac{d\underline{v}}{dt} = \underline{X}(\underline{v}) \ , \quad \underline{v} \text{ a curve in } P \ , \tag{2}$$

where $\underline{X}$ does not depend on time unless the boundary conditions or body force are time dependent. In this abstract notation the Navier-Stokes equations, boundary conditions, divergence-free condition, etc., look like an ordinary differential equation (1). In the present circumstances $\underline{X}$ is an unbounded operator. However, we can still apply the stability concepts developed previously for smooth $\underline{X}$'s , due to the following conditions:

1. Equation (2) really does determine the dynamics on P . This

follows from the short t-interval existence and uniqueness theorem for the Navier-Stokes equations (see [12] in the list of references in the Preface; here the technical conditions are precisely spelled out).

2. The stability theorem <u>really</u> works for the Navier-Stokes equations. This is a nontrivial matter (see Serrin [11] and the footnote below for additional references).

For (2), $\underline{DX}$ is obtained via (Fréchet) differentiation and $D\underline{X}(\underline{v}_0)$ is a linear differential operator. Its spectrum will consist of infinitely many eigenvalues or it may be continuous. For $\underline{v}_0$ to be stable the <u>entire</u> spectrum must lie in the left half-plane of $\mathbb{C}$ (cf. Figure 18-1).

<u>Examples</u>. Flow in a pipe and Couette flow are stable (in fact globally stable) if Re, the Reynolds number, is not too big. Couette flow is very instructive (see [11]).

We are interested in the stability of flows as Re is increased. In this regard we consider $\underline{X}$ to be parameterized by Re and study the behavior of the spectrum of $D\underline{X}(\underline{v}_0)$ as a function of this parameter. As Re is increased we anticipate that conjugate pairs of eigenvalues of $D\underline{X}(\underline{v}_0)$ may drift across the imaginary axis (Figure 18-3). In this case stability will be lost. Insight into what it will be replaced by is given by the

<u>Hopf Bifurcation Theorem</u>.* (Nontechnical version.) *Let* $\underline{X}_\mu$ *be a*

---

*References are J. Marsden and M. McCracken "The Hopf Bifurcation", Springer notes in Applied Mathematics (1976) and D. Sattinger, "Lectures on Stability and Bifurcation Theory," Springer Lecture Notes #309(1973). Books which approach this subject from a more physical point of view are C.C. Lin, "The Theory of Hydrodynamic Stability," Cambridge (1955) and Chandrasekar, "Hydrodynamic and Hydromagnetic Stability," Oxford (1961).

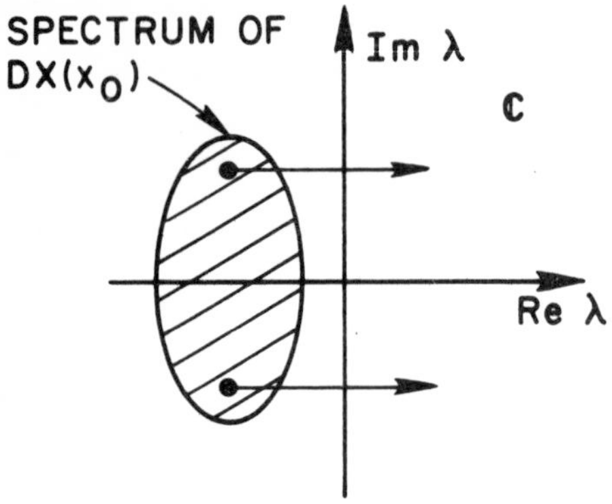

FIG. 18-3

parametrized vector field (parameter μ *) with fixed point* $\underline{x}_0$. *Assume the eigenvalues of* $D\underline{X}_\mu(\underline{x}_0)$ *are in the left half-plane of* $\mathbb{C}$ *for all* $\mu < \mu_0$, *where* μ_0 *is fixed. Assume that as* μ *is increased, a single conjugate pair* $\lambda($ *of eigenvalues crosses the imaginary axis with nonzero speed at* $\mu = \mu_0$ *(Fig. 18-4). Then there is a family of closed orbits of* $\underline{X}$ *near* μ_0 . *If the point* $\underline{x}_0$ *is stable for* $\underline{X}_{\mu_0}$, *then the closed orbits appear for* $\mu > \mu_0$ *and are stable. For each* $\mu > \mu_0$, *near* μ_0 , *there is one corresponding stable closed orbit. The period is approximately equal to* $\mathrm{Im}\,(\frac{\lambda(\mu_0)}{2\pi})$.

Roughly speaking, what this means is that when stability is lost, as above, a stable point is replaced by a stable closed orbit.

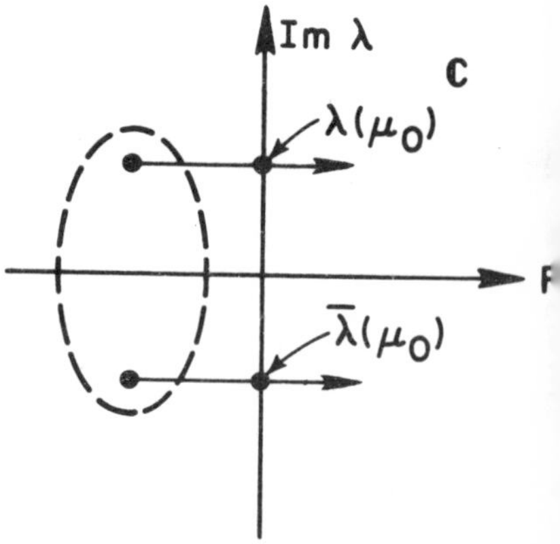

FIG. 18-4

Translated to the case of fluid mechanics this means that a stationary solution (fixed point) of the Navier-Stokes equations is replaced by a periodic solution (stable closed orbit). For example, consider the case of flow around a cylinder. As Re is increased, the stationary solution depicted in Figure 18-5 becomes unstable and goes to a stable periodic solution--the wiggly wake depicted in Figure 18-6.

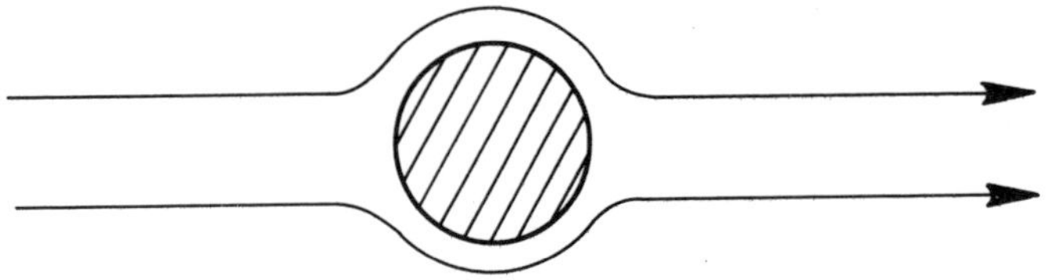

FIG. 18-5

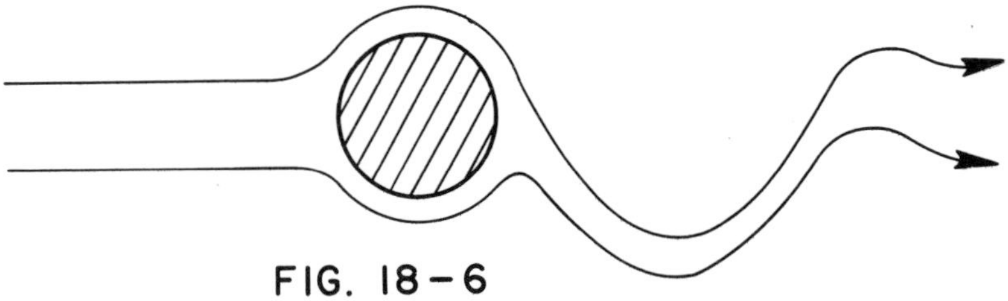

FIG. 18-6

We say that a <u>bifurcation</u> has occurred when Re hits the critical value where stability is lost and is replaced by oscillations. The term is used generally when sudden qualitative changes occur. In Hopf's theorem the stability analysis fails at μ_0 and stability must be determined by "higher-order" analysis; i.e., the stability theorem, based on first-order analysis just fails at $\mu = \mu_0$. How to do this is discussed in the cited references. One should also note that in highly symmetrical situations such as Couette flow, the hypotheses of Hopf's theorem (simplicity of the eigenvalues, or the nonzero speed assumption) can fail. Again further analysis in this situation is needed.

As Re is increased additional bifurcations occur and eventually turbulence ensues. We don't even attempt to define turbulence as this subject, despite tremendous interest and extensive literature, is not well understood. We shall return to make a few brief comments later in §23.

§19. A Brief Survey of Thermodynamics.

In order to properly understand compressible flow, it is necessary to clearly understand the thermodynamic assumptions underlying the theory. This section attempts to set these assumptions out clearly and concisely. In this section we shall make frequent use of the definitions and results of Sections 5-8.

Definition. Let h be a function depending on t, $\underline{x}$ and $\underline{n}$ and let e and r be functions depending on t and $\underline{x}$. We say $(\rho, \underline{v}, \underline{f}, \underline{t}, e, r, h)$ satisfy the balance of energy principle if for all regions W_t

$$\frac{d}{dt}\int_{W_t} \rho(e + \tfrac{1}{2}\,\underline{v}\cdot\underline{v})dx = \int_{W_t} \rho(\underline{f}\cdot\underline{v} + r)dx + \int_{S_t} (\underline{t}\cdot\underline{v} + h)da . \quad (1)$$

The physical jargon is that e represents the internal energy of a fluid, r the heat supply, and hda the heat flux across a surface with area element da oriented with normal $\underline{n}$. As before, $\underline{v}$ is the velocity of our medium and W_t is a moving volume, S_t being its surface.

Theorem. *The balance of energy principle implies that there exists a vector function* $\underline{q}(t, \underline{x})$ *such that* $h(t, \underline{x}, \underline{n}) = -\underline{q}(t, \underline{x})\cdot\underline{n}$.

Proof. The proof of this theorem proceeds along identical lines as that of Cauchy's Theorem; cf. §7. ∎

The vector function $\underline{q}$ is called the heat flux vector. Let

$$K = \int_{W_t} \tfrac{1}{2} \rho \, \underline{v} \cdot \underline{v} \, dx \, ,$$

$$E = \int_{W_t} \rho e \, dx \, ,$$

$$W = \int_{W_t} \rho \underline{f} \cdot \underline{v} \, dx + \int_{S_t} \underline{t} \cdot \underline{v} \, da \, ,$$

and

$$Q = \int_{W_t} \rho r \, dx + \int_{S_t} h \, da \, ,$$

be the kinetic energy, internal energy, mechanical power, non-mechanical power, respectively, of the volume W_t . Then (1) may be written as

$$\frac{d}{dt}(K + E) = W + Q \, , \tag{1'}$$

The validity of (1) (or (1')) is called the first law of thermodynamics for the volume of fluid W_t . In words, the rate-of-change of energy equals the work done plus heat supplied to the fluid.

Theorem. *Assume the integrands in* (1) *are* C^1 *and* W_t *is a nice subregion of* Ω . *Furthermore, assume the continuity equation, equations of motion, and Cauchy's second law hold. Then* (1) *implies*

$$\rho \frac{D}{Dt} e + \operatorname{div} \underline{q} = \operatorname{tr} \underline{T} \, \underline{D} + \rho r \, , \tag{2}$$

the local form of the first law of thermodynamics.

Sketch of proof. The proof of this result uses the Transport Theorem and Gauss' Theorem in the same way as the localization of the balance of momentum. The reader should carry out the details as an exercise. ■

An adiabatic process is one for which $\underline{q} = \underline{0}$.

Definition. Let η and θ be functions of t and $\underline{x}$. Assume $\theta(t, \underline{x}) > 0$ for all $t \in \mathbb{R}$ and $\underline{x} \in \Omega$. We say $(\rho, \eta, r, \theta, \underline{q})$ satisfy the entropy production inequality for W_t if

$$\frac{d}{dt} \int_{W_t} \rho \, \eta \, dx \geq \int_{W_t} \rho \frac{r}{\theta} \, dx - \int_{S_t} \frac{1}{\theta} \underline{q} \cdot \underline{n} \, da \; . \qquad (3)$$

Here η represents the entropy of the fluid and θ the (absolute) temperature. (3) is also referred to as the Clausius-Duhem inequality. We are tempted to describe (3) as the second law of thermodynamics, but just what the second law is seems ambiguous (see Truesdell's "Rational Thermodynamics," McGraw-Hill, (1969) for a discussion of the confusing state of affairs in classical thermodynamics and the modern continuum point of view).

Theorem. *Assume the integrands in* (3) *are* C^1 *and* W_t *is a nice subregion of* Ω . *In addition, assume the continuity equation holds. Then* (3) *implies the local form*

$$\rho \frac{D}{Dt} \eta \geq \rho \frac{r}{\theta} - \frac{1}{\theta} \operatorname{div} \underline{q} + \frac{1}{\theta^2} \underline{q} \cdot \underline{\nabla} \theta \; . \qquad (4)$$

This is proved by the now familiar localization procedure using the transport theorem.

(1) and (3) are our basic thermodynamical hypotheses. Note that for an adiabatic process in which there is no heat supply (4) implies $D\eta/Dt \geqslant 0$. An isentropic process is one for which $D\eta/Dt = 0$.

The free energy is defined by* $\psi = e - \theta\eta$. The effect the Clausius-Duhem inequality has on the structure of fluid dynamical theories is exemplified by the following theorem.

Theorem. *Assume the following constitutive equations hold* (i.e., ψ, η, $\underline{q}$, $\underline{T}$ *are functions of the variables shown*):

$$\psi = \hat{\psi}(V, \underline{D}, \theta, \underline{\nabla}\theta) ,$$

$$\eta = \hat{\eta}(V, \underline{D}, \theta, \underline{\nabla}\theta) ,$$

$$\underline{q} = -k\underline{\nabla}\theta , \qquad k = \hat{k}(V, \theta) ,$$

$$\underline{T} = -p\underline{I} + \underline{\tilde{T}} ,$$

where (cf. §16):

$$p = \hat{p}(V, \theta) ,$$

$$\underline{\tilde{T}} = \lambda(\mathrm{tr}\,\underline{D})\underline{I} + 2\mu\underline{D} , \qquad \lambda = \hat{\lambda}(V, \theta) , \quad \mu = \hat{\mu}(V, \theta) ,$$

and $V = 1/\rho$ *is the specific volume. Furthermore, assume* $\hat{\psi}$, $\hat{\eta}$, $\hat{k}$, $\hat{p}$

*Those in mechanics will recognize a Legendre transformation here.

$\hat{\lambda}$ *and* $\hat{\mu}$ *are* C^1 *functions. Then* (2), (4) *and the continuity equation imply that*

$$\hat{\psi} = \hat{\psi}(V, \theta) ,$$

$$\hat{\eta} = -\frac{\partial \hat{\psi}}{\partial \theta} ,$$

$$\hat{p} = -\frac{\partial \hat{\psi}}{\partial V}$$

and

$$\hat{k} \geqslant 0 , \quad \hat{\mu} \geqslant 0 , \quad \hat{\lambda} + \frac{2}{3}\hat{\mu} \geqslant 0 .$$

Proof. We combine (2) and (4) by eliminating r to obtain the reduced dissipation inequality

$$\rho\left(\frac{D\psi}{Dt} + \eta\,\frac{D\theta}{Dt}\right) - \mathrm{tr}\,\underline{T}\,\underline{D} + \frac{1}{\theta}\,\underline{q}\cdot\underline{\nabla}\theta \leqslant 0 .$$

Substituting the constitutive equations into the above yields

$$\rho\left\{\left(\hat{p} + \frac{\partial\hat{\psi}}{\partial V}\right)\frac{DV}{Dt} + \mathrm{tr}\,\frac{\partial\hat{\psi}}{\partial \underline{D}}\,\frac{D}{Dt}\,\underline{D} + \left(\eta + \frac{\partial\hat{\psi}}{\partial\theta}\right)\frac{D\theta}{Dt} + \frac{\partial\hat{\psi}}{\partial\underline{\nabla}\theta}\cdot\frac{D}{Dt}\,\underline{\nabla}\theta\right\} \qquad (5)$$

$$- \hat{\lambda}(\mathrm{tr}\,\underline{D})^2 - 2\hat{\mu}\,\mathrm{tr}\,\underline{D}^2 - \frac{1}{\theta}\,\hat{k}\underline{\nabla}\theta\cdot\underline{\nabla}\theta \leqslant 0 ,$$

where we have used the continuity equation in the form $V\,\mathrm{tr}\,\underline{D} = DV/Dt$. The inequality (5) is to hold for all values of the arguments in the domains of the constitutive functions. Thus we are free to construct processes involving arbitrary values of $\underline{D}$, $D\,\underline{D}/Dt$, DV/Dt , $D\theta/Dt$

and $D\underline{\nabla}\theta/Dt$ since we can always satisfy the equations of motion and energy by suitable choices of $\underline{f}$ and r. For example consider a process in which $D\theta/Dt$ is made arbitrarily large (positive), then consider everything the same except specify $D\theta/Dt$ arbitrarily negative. Since $D\theta/Dt$ only occurs in one place in (5), its coefficient must vanish identically. Thus $\eta = -\partial\hat{\psi}/\partial\theta$. Similar reasoning yields $\partial\hat{\psi}/\partial\underline{D} = 0$ and $\partial\hat{\psi}/\partial\underline{\nabla}\theta = 0$. Thus $\hat{\psi} = \hat{\psi}(V, \theta)$. Now assume $\underline{\nabla}\theta = \underline{0}$. Pick $\underline{D}$ arbitrarily small (in norm) so that DV/Dt does not vanish. In this case the terms of second order in $\underline{D}$ may be neglected compared with the first term in (5). Reasoning as before, we conclude $\hat{p} = -\partial\hat{\psi}/\partial V$. Now assume $\underline{\nabla}\theta \neq \underline{0}$ and $\underline{D} = \underline{0}$. It follows that $\hat{k} \geq 0$. Take $\underline{\nabla}\theta = \underline{0}$, $\mathrm{tr}\,\underline{D} = 0$ and $\mathrm{tr}\,\underline{D}^2 \neq 0$. It follows that $\hat{\mu} \geq 0$. Finally, take $\underline{\nabla}\theta = \underline{0}$ and $\underline{D} = \underline{I}$ to get $\hat{\lambda} + 2\hat{\mu}/3 \geq 0$. ■

The constitutive equation $\underline{q} = -k\underline{\nabla}\theta$ is often referred to as Fourier's law of heat conduction; k is the conductivity. One can actually prove the same theorem as above with weaker assumptions, but this is outside our main interest here.

Corollary. *Let* $\hat{e}(V, \theta, \eta) = \hat{\psi}(V, \theta) + \theta\eta$. *Then* $\hat{e} = \hat{e}(V, \eta)$, $\hat{p} = -\partial\hat{e}/\partial V$ *and* $\theta = \partial\hat{e}/\partial\eta$.

Proof. Differentiating $\hat{e}$ with respect to its arguments, we obtain

$$\frac{\partial\hat{e}}{\partial V} = \frac{\partial\hat{\psi}}{\partial V},$$

$$\frac{\partial\hat{e}}{\partial\theta} = \frac{\partial\hat{\psi}}{\partial\theta} + \eta = 0,$$

and

$$\frac{\partial\hat{e}}{\partial\eta} = \theta.$$

The conclusions are immediate from the results of the theorem. ■

The roots and physical basis of thermodynamics are hard to get hold of and we have not attempted such a task here. The basic mathematical structure however, is simple and clean. For discussion of the physical basis of the laws, especially the elusive second law, the reader is referred to any thermodynamics book (one should not have high expectations in such a venture).

In subsequent sections we shall see how useful thermodynamic ideas are in formulating the equations of compressible flow, such as the flow of an ideal gas.

§20. Ideal Compressible Flow.

In the present section we assume the fluid in question is inviscid and the flow is adiabatic. Under these assumptions the momentum and energy balances become

$$\rho \frac{D\underline{v}}{Dt} = -\text{grad } p \quad , \tag{1}$$

and

$$r = \frac{De}{Dt} + \rho \frac{DV}{Dt} \; , \tag{2}$$

respectively. In arriving at (2) we have employed the continuity equation and equation (3) of Section 19. Throughout this section we make use of the thermodynamic relationships derived in the previous section. A summary of those relations and the definitions of some new constitutive functions are as follows:

$$\left.\begin{aligned}
\eta &= \hat{\eta}(V, \theta) = -\frac{\partial\hat{\psi}}{\partial\theta}(V, \theta) \; , \\
\theta &= \hat{\theta}(V, \eta) = \frac{\partial\hat{e}}{\partial\eta}(V, \eta) \; , \\
p &= \hat{p}(V, \theta) = -\frac{\partial\hat{\psi}}{\partial V}(V, \theta) \; , \\
&= \hat{\hat{p}}(V, \eta) = \hat{p}(V, \hat{\theta}(V, \eta)) = -\frac{\partial\hat{e}}{\partial V}(V, \eta) \; , \\
V &= \hat{V}(p, \theta) \; ,
\end{aligned}\right\} \tag{3}$$

$$e = \hat{\hat{e}}(V,\ \theta) = \hat{e}(V,\ \hat{\eta}(V,\ \theta))\ ,$$

$$= \hat{\hat{\hat{e}}}(p,\ \theta) = \hat{\hat{e}}(\hat{V}(p,\ \theta),\ \theta)\ .$$

The function $\hat{V}$ is obtained by inverting the relation $p = \hat{p}(V,\ \theta)$. (3) can be used to obtain a simplified version of (2), viz.,

$$r = \frac{De}{Dt} + p\,\frac{DV}{Dt} = \left(\frac{\partial\hat{e}}{\partial V} + p\right)\frac{DV}{Dt} + \theta\,\frac{D\eta}{Dt} = \theta\,\frac{D\eta}{Dt}\ . \tag{4}$$

From this relation we see that if there is no heat supply ($r = 0$) , then the flow is isentropic, i.e., the entropy of each fluid particle remains constant. This does not mean that the entropy is spatially constant; if it is constant the flow is called <u>homentropic</u> (note: some writers use the term isentropic for the spatially constant case).

Equations (1), (4), the continuity equation, and the constitutive equations $\eta = \hat{\eta}(V,\ \theta)$ and $p = \hat{p}(V,\ \theta)$ constitute a formally deterministic system of equations for adiabatic, inviscid flow, i.e., there are seven equations for determining the seven unknowns $(V,\ \underline{v},\ p,\ \theta,\ \eta)$. The boundary conditions for these equations are the same as those for the equations of an ideal fluid (cf. §9). The initial conditions are ρ, $\underline{v}$ and η specified at $t = 0$. To make this system explicit the functions $\hat{\eta}$ and $\hat{p}$ must be specified. Our objective for the remainder of this section is to obtain specific forms for $\hat{p}$ and $\hat{\eta}$ for a case of broad practical significance.

Employing (3), equation (2) can be expressed in the alternative forms

$$r = \left\{\frac{\partial \hat{\hat{e}}}{\partial V} + p\right\} \frac{DV}{Dt} + \frac{\partial \hat{\hat{e}}}{\partial \theta} \frac{D\theta}{Dt} \, , \tag{5}$$

and

$$r = \left\{\frac{\partial \hat{\hat{\hat{e}}}}{\partial \theta} + p \, \frac{\partial \hat{V}}{\partial \theta}\right\} \frac{D\theta}{Dt} + \left\{\frac{\partial \hat{\hat{e}}}{\partial p} + p \, \frac{\partial \hat{V}}{\partial p}\right\} \frac{Dp}{Dt} \, . \tag{6}$$

(5) and (6) are used to define the "specific heats" of a fluid. Namely, the heat required to raise the temperature of a unit mass of fluid by one unit at constant volume, in one unit of time, is (set $DV/Dt = 0$ in (5))

$$C_V = \frac{\partial \hat{\hat{e}}}{\partial \theta} \qquad (\, = \underline{\text{specific heat at constant volume}}) \, ,$$

and the heat required to raise the temperature of a unit mass of fluid one unit at constant pressure, in one unit of time, is (set $Dp/Dt = 0$ in (6))

$$C_p = \frac{\partial \hat{\hat{\hat{e}}}}{\partial \theta} + p \, \frac{\partial V}{\partial \theta} \qquad (\, = \underline{\text{specific heat at constant pressure}}) \, .$$

Since $\dfrac{\partial \hat{\hat{\hat{e}}}}{\partial \theta} = \dfrac{\partial \hat{\hat{e}}}{\partial V} \dfrac{\partial \hat{V}}{\partial \theta} + \dfrac{\partial \hat{\hat{e}}}{\partial \theta}$, the specific heat at constant pressure can also be written as

$$C_p = C_V + \left\{p + \frac{\partial \hat{\hat{e}}}{\partial V}\right\} \frac{\partial \hat{V}}{\partial \theta} \, .$$

An <u>ideal gas</u> is defined to be one for which

$$pV = R\theta \ , \tag{7}$$

where R is a constant. (7) is sometimes referred to as Boyle's law. In the present circumstances it may be viewed as a constitutive hypothesis providing explicit forms for $\hat{p}$ and $\hat{V}$, i.e., $p = \hat{p}(V, \theta) = R\theta/V$ and $V = \hat{V}(p, \theta) = R\theta/p$. It remains to obtain an expression for $\hat{\eta}$.

Theorem. *Assume equations* (3) *hold for an ideal gas. Then* $e = \hat{\hat{e}}(V, \theta) = \hat{\hat{e}}(\theta)$, *i.e.,* e *is a function of* θ *only.*

Proof. Combining (3) and (7) results in $-\partial\hat{\psi}/\partial V = p = R\theta/V$. Integrating $V\partial\hat{\psi}/\partial V = -R\theta$ yields $\hat{\psi} = -R\theta \log V + g(\theta)$, where g is an arbitrary function of θ. Since $e = \psi + \theta\eta$, by definition of ψ, (3) and the previous result implies $\hat{\hat{e}} = \hat{\psi} - \theta\partial\hat{\psi}/\partial\theta = g(\theta) - \theta g'(\theta)$. ■

Corollary. $C_V = -\theta g''(\theta)$ *and* $C_p = C_V + R$.

Proof. The expression for C_V follows immediately from the definition $C_V = \partial\hat{\hat{e}}/\partial\theta$ and the theorem. The definition for $\hat{V}$ provided by (7), the definition $C_p = C_V + \{p + \partial\hat{\hat{e}}/\partial V\}\partial\hat{V}/\partial\theta$, and the theorem yield the second result. ■

We observe from the corollary that C_V and C_p are functions of θ only. This may be used to obtain an expression for $\hat{\hat{e}}$ given C_V, i.e., $\hat{\hat{\varepsilon}}(\theta) = \int^{\theta} C_V(s)ds$. We can also obtain an expression for $\hat{\eta}$, i.e. $\hat{\eta} = -\partial\hat{\psi}/\partial\theta = -R \log V + g' = -R \log V + \int^{\theta} \frac{C_V(s)}{s} ds$. In addition, if C_V is constant then C_p is also. Let $\gamma = C_p/C_V$. Then $R = C_V(\gamma - 1)$.

Theorem. *Assume that equations* (3) *hold for an ideal gas for which* C_V *is constant. Then* $\hat{\hat{e}} = C_V\theta$ + *const. and* $\hat{\eta} = \frac{R}{(\gamma-1)}\log(\theta V^{(\gamma-1)})$ + *const.*

Proof. That $\hat{\hat{e}} = C_V$ + const. is immediate from the remarks preceding the theorem. Also from these remarks we have

$$\hat{\eta} = -R \log V + C_V \log \theta + \text{const.}$$

Substituting $C_V = R/(\gamma - 1)$ then yields the result. ∎

Corollary. *Assume (without loss of generality) that the constant in the expression for* $\hat{\eta}$ *is zero. Then*

$$p = R\rho^{\gamma} \exp(\eta/C_V) .$$

Proof. Substitute p/R for θ/V in the expression for $\hat{\eta}$ and solve. ∎

Remarks. The relation $e = C_V\theta$ is taken to be the defining property of a polytropic gas. The result of the corollary is often combined with

(1), the continuity equation, and the isentropic flow condition $D\eta/Dt = 0$, to form a formally deterministic system of six equations in the six unknowns $(\rho, \underline{v}, p, \eta)$. These are the equations of isentropic, adiabatic, inviscid flow of a fluid with C_V constant.

§21. Shock Waves.

To study the phenomenon of shock waves in a compressible gas we will make some drastic simplifying assumptions; namely:

1. The flow is one-dimensional, i.e.,

$$v = v_1(t, x) ,$$
$$v_2 = v_3 = 0 ,$$
$$p = p(t, x) ,$$

and

$$\rho = \rho(t, x) .$$

2. There are no viscous effects. Thus the constitutive equation is that for a perfect fluid (cf. §9).

3. The pressure is a given function of the density only, i.e., $p = \hat{p}(\rho)$, where $\hat{p}$ is smooth and $\hat{p}' = d\hat{p}/d\rho > 0$. The latter condition, as we shall see later, guarantees real wave speeds. (The condition that pressure is a function of density alone is discussed later.)

With the above assumptions, the continuity equation and x-direction momentum equation can be written

$$\left.\begin{aligned} &\frac{\partial\rho}{\partial t} + \frac{\partial\rho v}{\partial x} = 0 \\ \text{and} \\ &\frac{\partial\rho v}{\partial t} + \frac{\partial\rho v^2}{\partial x} + \frac{\partial\hat{p}}{\partial x} = 0 \end{aligned}\right\} \qquad (1)$$

respectively. It is implicitly assumed that ρ and v are C^1 functions of x and t, so that (1) makes sense. Thus it is equally valid to write

$$\left.\begin{aligned} &\frac{\partial\rho}{\partial t} + \rho\frac{\partial v}{\partial x} + v\frac{\partial\rho}{\partial x} = 0 \\ \text{and} \\ &\frac{\partial v}{\partial t} + v\frac{\partial v}{\partial x} + \frac{1}{\rho}\frac{\partial\hat{p}}{\partial x} = 0 \end{aligned}\right\} \qquad (1)'$$

in place of $(1)_1$ and $(1)_2$, respectively (note that we have combined $(1)_1$ and $(1)_2$ to obtain $(1)'_2$).

To investigate the properties of these equations we will momentarily detour and consider linear first-order systems of the form

$$\frac{\partial\underline{u}}{\partial t} = \underline{A}\,\frac{\partial\underline{u}}{\partial x}\,, \qquad (2)$$

where $\underline{u} = \underline{u}(x, t)$ is an n-vector and $\underline{A}$ is a constant $n \times n$ matrix. For the simple case in which $n = 1$, (2) reduces to

$$\frac{\partial u}{\partial t} = a\frac{\partial u}{\partial x}\,, \quad a \text{ const.}, \qquad (3)$$

which has the general solution

$$u(t, x) = f(x + at)\,,$$

where $u(x, 0) = f(x)$ is a given function. The physical interpretation of this solution is that the initial data, $f(x)$, is propagated to the left (negative x-direction) with velocity a, without incurring distortion. One frequently refers to a solution such as this as a wave (in the present circumstances this terminology is meant to be heuristic and not a precise definition). Note that the solution is constant on the lines $x + at = \text{const.}$ in the $x\,t$-plane.

We are interested in the case in which $\underline{A}$ is diagonalizable over the reals; i.e., possesses the property that there exists a nonsingular $n \times n$ matrix $\underline{T}$ such that

$$\underline{T}^{-1}\,\underline{A}\,\underline{T} = \underline{\Lambda} = \begin{pmatrix} \lambda_1 & 0 & \cdots & 0 \\ 0 & \lambda_2 & & \vdots \\ \vdots & & \ddots & 0 \\ 0 & \cdots & 0 & \lambda_n \end{pmatrix}$$

where λ_i is real, $i = 1, \ldots, n$. The λ_i's are the eigenvalues of $\underline{A}$. When $\underline{A}$ possesses this property the system (2) is said to be hyperbolic. A special type occurs when $\underline{A}$ is symmetric. Then we have a symmetric hyperbolic equation. If (2) is hyperbolic, it can be converted to diagonal form, namely

$$\frac{\partial \underline{w}}{\partial t} = \underline{\Lambda}\,\frac{\partial \underline{w}}{\partial x}\,,$$

where $\underline{w} = \underline{T}^{-1}\underline{u}$. Thus the problem of solving a system of the form (2) which is hyperbolic, reduces to solving single equations for the components of $\underline{w}$. These equations have the form (3) with a replaced by λ_i , and thus their solutions are waves propagating with velocities λ_i . Each one of these waves is constant on the lines $x + \lambda_i t = \text{const.}$; these lines are called <u>characteristics</u>.

For the case in which the matrix $\underline{A}$ is not constant, but a function of $\underline{u}$, x and t , a similar, but slightly more complicated situation occurs. The fact that $\underline{A}$ may depend on $\underline{u}$ means that the equations are nonlinear. Only for nonlinear equations can "shocks" develop. The existence theory for such equations is a subject of current research by many authors,* although some information can be found in Courant-Hilbert, "Methods of Mathematical Physics," Vol II Interscience (1962).

(1)′ can be put in the form (2) where $\underline{A}$ is a function of $\underline{u} = (\rho, v)$, viz.,

$$\frac{\partial}{\partial t}\begin{pmatrix}\rho \\ v\end{pmatrix} = -\begin{pmatrix} v & \rho \\ c^2/\rho & v\end{pmatrix}\frac{\partial}{\partial x}\begin{pmatrix}\rho \\ v\end{pmatrix},$$

where $c = (\hat{p}')^{\frac{1}{2}}$; c is called the <u>speed of sound</u> relative to the fluid. Here the coefficient matrix is diagonalizable for all values of ρ and v , and the eigenvalues are $v \pm c$, which are real by virtue of assumption 3. The characteristics for this case are the curves obtained by integrating the equations $dx/dt = v \pm c$. Along these characteristics the <u>Riemann invariants</u>

---

*See for example, P. Lax, "Hyperbolic Systems of Conservation Laws and the Mathematical Theory of Shock Waves," SIAM (1973) <u>28</u> (1972) 1-38, T. Kato, Arch. Rat. Mech. An. <u>58</u> (1975) 181-205 and T. Hughes, T. Kato and J. Marsden (preprint).

$$r_{\pm} = v \pm \int^{\rho} \frac{c(\bar{\rho})}{\bar{\rho}} \, d\bar{\rho} ,$$

are constant. This amounts to a simple computation, which the reader should verify (hint: differentiation along the characteristics is facilitated by using the directional derivative $\partial/\partial t + (v \pm c)\, \partial/\partial x$). The solution can be constructed from the Riemann invariants with the help of numerical methods.* However, a complication can arise. Consider in particular the model equation

$$\frac{\partial v}{\partial t} = -v \frac{\partial v}{\partial x} = -\frac{\partial}{\partial x}\left(\frac{v^2}{2}\right) .$$

The solution of this equation is given implicitly by $v = f(x - vt)$ where $v(x, 0) = f(x)$ is the initial data. From this we infer that v is constant along lines $x - vt = \text{const}$. But if $f(x)$ is a nonconstant function, this may lead to a contradiction; namely v must take on two different values at a single point (Figure 21-1 illustrates the dilemma). Thus in the nonlinear case where the coefficient matrix is not constant, characteristics can intersect, which contradicts the constancy of the Riemann invariants. Something has gone awry--but what?

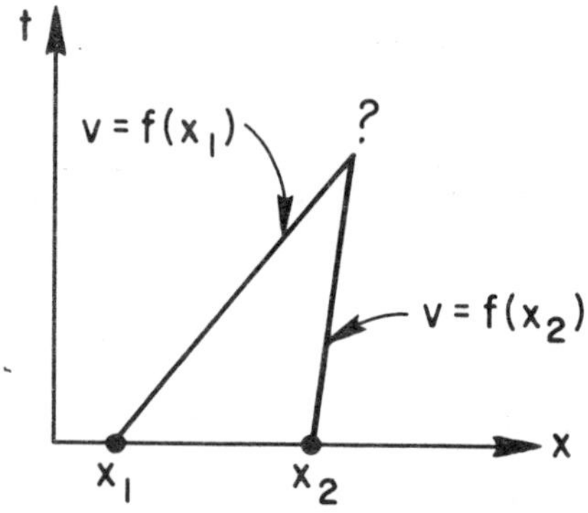

FIG. 21-1 INTERSECTION OF CHARACTERISTICS FOR THE MODEL EQUATION

One postulate is that a discontinuity will appear instead of

*Numerical analysis is an important and active area of research in fluid mechanics; numerical work of Lax, Chorin and many others has useful practical and theoretical applications.

the solution becoming multivalued. The discontinuity--called a <u>shock</u>--will then separate the regions in which the different values are assumed. At a discontinuity the differential equations (1) become meaningless. However, the basic balance conditions (i.e., conservation of mass, balance of momentum and energy) cannot be violated. The integral forms of the balance laws impose restrictions on the behavior of the shock wave. The problem then is to incorporate into a meaningful picture the following conditions:

(i) The solution of $(1)'$ is composed of regions, separated by shocks, in which ρ and v are smooth functions.

(ii) Where the solution is smooth the differential equations $(1)'$ are to hold in the classical sense.

(iii) At a shock the integral balance laws remain valid.

(iv) The whole picture must remain physically meaningful when viscosity and thermodynamic effects are accounted for.

A nice way of synthesizing (i)-(iv) is through the theory of weak solutions which we shall now describe.

Let $\underline{f} = (f_1, f_2)$ be a vector-valued function of $\underline{u}$, x and t. If $\underline{f}$ and $\underline{u}$ are smooth, it is permissible to consider equations of the form

$$\text{DIV } \underline{f} \overset{\text{def.}}{=} \frac{\partial f_1}{\partial t} + \frac{\partial f_2}{\partial x} = 0 ,$$

which is said to be in (differential) <u>conservation</u> <u>form</u>. Equations $(1)_1$ and $(1)_2$ can be put in conservation form, viz., let

$$\underline{f}^1 = (\rho, \rho v) \quad \text{and} \quad \underline{f}^2 = (\rho v, \rho v^2 + \hat{p}) \ ; \tag{4}$$

then DIV $\underline{f}^1 = 0$ and DIV $\underline{f}^2 = 0$ are in conservation form, and are equivalent to $(1)_1$ and $(1)_2$, respectively.

Definition. A test function $\psi(x, t)$ is a C^∞ function with compact support (i.e., it is zero outside a bounded set in the plane).

Lemma. Suppose $\underline{f}(\underline{u})$ and $\underline{u}(x, t)$ are smooth functions. Then DIV $\underline{f}$ = 0 if and only if $(\psi, \text{DIV } \underline{f}) \overset{\text{def.}}{=} \int_{\mathbb{R}^2} \psi \text{ DIV } \underline{f} \, dx \, dt = 0$ for all test functions ψ .

Proof. Clearly DIV $\underline{f} = 0$ implies $(\psi, \text{DIV } \underline{f}) = 0$ for all test functions ψ . For the other direction we will assume the contrary and deduce a contradiction. Assume $(\psi, \text{DIV } \underline{f}) = 0$ for all test functions ψ , but DIV $\underline{f} \neq 0$ at a point (x_1, t_1) . By continuity, there exists an open ball around (x_1, t_1) such that DIV $\underline{f}$ does not change sign. Pick ψ such that its support is contained in this ball, it is $\geqslant 0$ everywhere, and identically equal to 1 at (x_1, t_1) . This implies $(\psi, \text{DIV } \underline{f}) > 0$, which is a contradiction. ∎

Thus as long as $\underline{f}$ and $\underline{u}$ are smooth DIV $\underline{f} = 0$ and $(\psi, \text{DIV } \underline{f})$ = 0 for all test functions ψ are equivalent conditions. The smoothness of $\underline{f}$ and $\underline{u}$ permits integration by parts:

$$(\psi, \text{DIV } \underline{f}) = -(\text{GRAD } \psi, \underline{f}) \ .$$

Thus DIV $\underline{f} = 0$ is equivalent to $(\text{GRAD } \psi, \underline{f}) = 0$ for all test functions ψ. Note that this condition involves no differentiations of $\underline{f}$ and can be given meaning even if $\underline{f}$ is discontinuous. $(\text{GRAD } \psi, \underline{f}) = 0$ is called the <u>weak form</u> of the equation DIV $\underline{f} = 0$. A function $\underline{u}$ which satisfies $(\text{GRAD } \psi, \underline{f}(\underline{u})) = 0$, for all test functions ψ, is called a <u>weak solution</u> of DIV $\underline{f} = 0$. If $\underline{u}$ is a C^1 function of x and t then it is also a <u>strong solution</u> (i.e., it satisfies DIV $\underline{f} = 0$ in the classical sense). Clearly all strong solutions are weak solutions. However, the converse is not true. In the presence of shocks we take the weak form of the equation as fundamental.

Suppose $\underline{u}$, and consequently $\underline{f}$, are discontinuous across a smooth curve Σ in the x,t-plane, but elsewhere are smooth (see Figure 21-2). Let $\underline{n} = \underline{n}^+ = -\underline{n}^-$ be a normal vector field to Σ and denote by $\underline{f}^+$ the restriction of $\underline{f}$ to the part of the plane into which $\underline{n}^+$ points, etc. Then the weak form $(\text{GRAD } \psi, \underline{f}) = 0$ implies DIV $\underline{f} = 0$ on $\mathbb{R}^2 - \Sigma$ and $(\underline{f}^+ - \underline{f}^-)\cdot\underline{n} = 0$ across Σ. This result can be obtained by integration by parts. The condition $(\underline{f}^+ - \underline{f}^-)\cdot\underline{n} = 0$ is called the <u>jump condition</u>. A more convenient notation is to

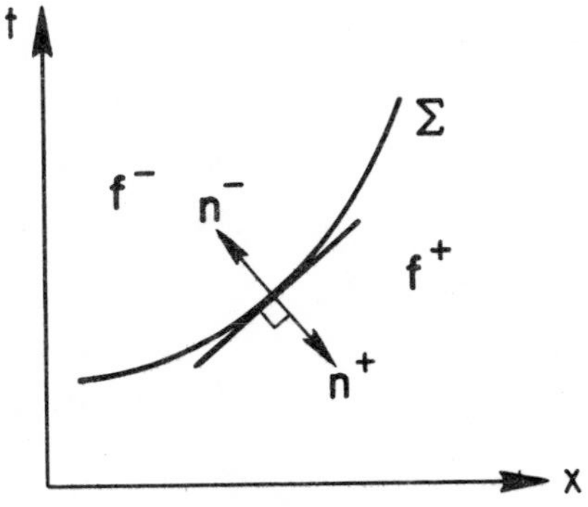

FIG. 21-2

use the jump operator $[f] \stackrel{\text{def.}}{=} \underline{f}^+ - \underline{f}^-$; then $[\underline{f}] \cdot \underline{n} = 0$. $\underline{n}$ can be taken to be $(s, -1)$, where s is the shock speed. The jump conditions corresponding to (4) are

$$\left.\begin{aligned} s[\rho] &= [\rho v] \\ \text{and} \qquad s[\rho v] &= [\rho v^2 + \hat{p}] \end{aligned}\right\} \qquad (5)$$

which manifest the conservation of mass and balance of momentum, respectively, across the shock.

Let us summarize what we have done. Anticipating solutions of (1) to possess discontinuities we have replaced (1) by corresponding weak forms

$$\left.\begin{aligned} &\int_{\mathbb{R}^2} \left(\frac{\partial\psi}{\partial t}\rho + \frac{\partial\psi}{\partial x}\rho v\right)dx\,dt = 0 \\ \text{and} \qquad &\int_{\mathbb{R}^2} \left(\frac{\partial\psi}{\partial t}\rho v + \frac{\partial\psi}{\partial x}(\rho v^2 + \hat{p})\right)dx\,dt = 0\,. \end{aligned}\right\} \qquad (6)$$

Where ρ and v are smooth, (6) is equivalent to (1). However, where ρ and v experience jumps, (6) implies that (5) holds.

If a viscous term $\nu \dfrac{\partial^2 v}{\partial x^2}$ is added to $(1)_2$, $(6)_2$ can be replaced by the weak form

$$\int_{\mathbb{R}^2} \left(\frac{\partial\psi}{\partial t}\rho v + \frac{\partial\psi}{\partial x}(\rho v^2 + \hat{p}) + \frac{\partial^2\psi}{\partial x^2}\nu v\right)dx\,dt\,. \qquad (7)$$

It can be shown that weak solutions of $(6)_1$ and (7) do not possess shocks.* Thus the effect of viscosity is to smooth out shock waves. It is formally clear however, that as $\nu \to 0$, weak solutions of $(6)_1$ and (7) coincide with those of $(6)_1$ and $(6)_2$.

There is still one more difficulty with (6). The class of weak solutions is too big, i.e., nonuniqueness can be shown to hold; see preceding footnote. The difficulty can be seen by considering the model equation

$$\frac{\partial v}{\partial t} + \frac{\partial}{\partial x}\left(\frac{v^2}{2}\right) = 0 . \tag{8}$$

The weak form corresponding to (8) is

$$\int_{\mathbb{R}^2} \left(\frac{\partial \psi}{\partial t} v + \frac{\partial \psi}{\partial x}\frac{v^2}{2}\right) dx\, dt = 0 . \tag{9}$$

The jump condition implied by (9) is

$$s[v] = \left[\frac{v^2}{2}\right] .$$

Suppose at time $t = 0$, v is given by

$$v(0, x) = \begin{cases} 0 & x < 0 \\ 1 & x > 0 \end{cases}$$

---

*See T.J.R. Hughes, "A Study of the One-Dimensional Theory of Arterial Pulse Propagation," SESM Report No. 74-13, University of California, Berkeley, 1974. Available from National Technical Information Service, Springfield, Virginia 22151, Accession No. PB-238 968/AS.

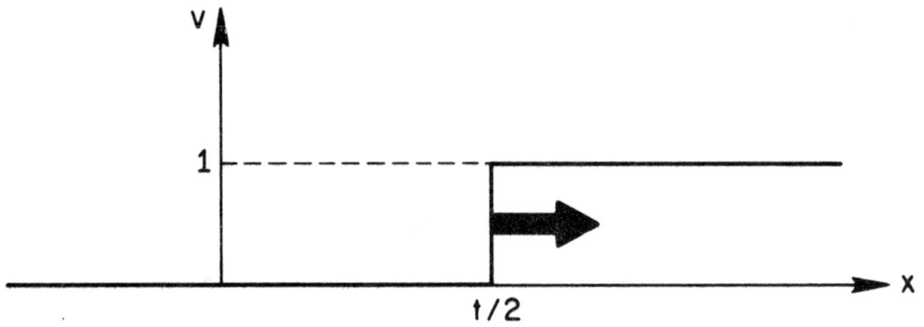

FIG. 21-3

The function (Figure 21.3)

$$v_I(t, x) = \begin{cases} 0 & x < t/x \\ 1 & x > t/2 \end{cases}$$

is a weak solution of (9) corresponding to the given initial data, but so is the function (Figure 21.4),

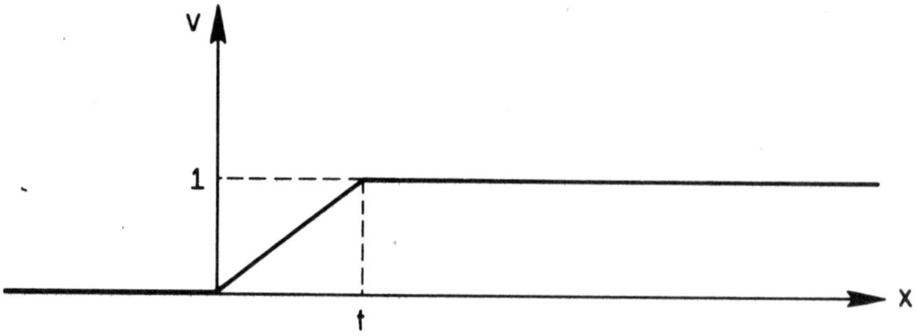

FIG. 21-4

$$v_{II}(x, t) = \begin{cases} 0 & x \leqslant 0 \\ x/t & 0 < x \leqslant t \\ 1 & t \leqslant x \end{cases}$$

which is continuous for all $t > 0$.

Much effort has been exerted to develop a criterion for selecting the physically relevant solution. For the model equation all criteria proposed are equivalent to the condition

$$v^- > s > v^+ , \tag{10}$$

hold on a line of discontinuity, where, as before, v^- and v^+ are the values of v to the left and right, respectively, of the discontinuity. The function v_{II} automatically satisfies this condition since it is continuous for $t > 0$. For v_I, $v_I^- = 0$, $v_I^+ = 1$ and $s = \frac{1}{2}$ which is in violation of (10). Thus v_{II} is the physically relevant solution.

For (6) one imposes the <u>entropy</u> (or shock admissibility) <u>condition</u>, which takes the forms:

$$(1) \quad \begin{cases} v^+ + c^+ < s < v^- + c^- \\ v^- - c^- < s \end{cases}$$

$$(2) \quad \begin{cases} v^+ - c^+ < s < v^- - c^- \\ s < v^+ + c^+ . \end{cases}$$

If either (1) or (2) is satisfied the shock in question is deemed admissible. Invoking the entropy condition leads to a definition of a unique weak solution for (6).*

The mathematical content of the entropy condition in the present setting is the requirement that discontinuities in the solution be allowed only where absolutely necessary; this necessity arises of course whenever characteristics would intersect if the shock were not interposed. The entropy condition excludes from consideration discontinuous solutions which satisfy the weak form of the equations in situations where the original equations (1) have smooth solutions.

The entropy condition does however have a deeper physical meaning. The theory presented in this section does not apply to real gas dynamics, except in very special situations. (It does, however, have an interesting application in the theory of water waves, where the equations of motion of the surface separating water and air can be written in a form very similar to (1).) The missing ingredient is the principle of conservation of energy. This principle leads to a differential equation which must be considered at the same time as the two equations (1). The resulting analysis proceeds as above, except that three conditions must be satisfied across the shock. This cannot be done if p is a function of ρ alone; p must be allowed to depend on another thermodynamical variable, e.g., the internal energy, and it must be allowed to be discontinuous across the shock. An entropy condition must be imposed in this more general setting, and here this condition serves a

---

*See P. Lax, "Hyperbolic Systems of Conservation Laws and the Mathematical Theory of Shock Waves," SIAM (1973) and references to work of J. Glimm, S. Krushkov and B. Quinn therein. Recent work has also been done by M. Crandall. (Isreal J. Math. 12(1972) 108-132.)

double function: It excludes mathematically "unnecessary" discontinuities, and it also reminds us that the assumption that dissipation and heat conduction are negligible is of course untenable in the presence of infinite velocity and temperature gradients. The shocks do have some (very small) thickness, in which an irreversible process is occurring. In this process entropy must increase. The entropy condition is in fact identical in the present situation to the requirement that the entropy increase. The entropy condition can be shown to have many interesting consequences; in particular it allows only compression shocks, i.e., shocks which have the property that the fluid crossing them finds itself with a higher pressure than before.

Studies of shock wave propagation in fluids is an active area of mathematical research and there are many basic unsolved problems, the most obvious of which is: What to do in three-dimensional space?

Problem. Show that for a gas for which $p = R\,\rho^{\gamma} \exp(\eta/C_V)$, $c^2 = \gamma R\theta$; cf. §20.

Problem. Show that shocks as described in this section cannot occur in an incompressible fluid.

§22. Boundary Layer Theory.

As was noted in §18, solutions of the Navier-Stokes equations are not close to solutions of the Euler equations for large Reynolds numbers Re (or small viscosities ν) in case boundaries are present. In 1904, Prandtl introduced boundary layer theory and the no-slip boundary condition for the Navier-Stokes equations to explain this.* Previously there was a great deal of confusion on the matter of large Reynolds numbers.

Prandtl inferred from the no-slip boundary condition that there would be a region of fluid near the boundary in which the viscous terms were as important as the inertial (= nonlinear) terms. Far from the boundary the flow would be driven primarily by the Euler equations. The region near the boundary is called the boundary layer. Prandtl derived the boundary layer equations, whose solution he believed approximated the solutions of the Navier-Stokes equations in the boundary layer. These boundary layer equations or modifications of them are in common use (see Figure 22-1).

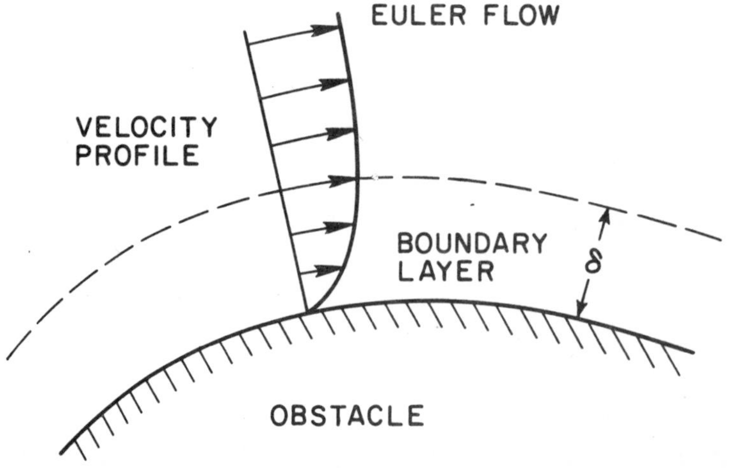

FIG. 22-1

*Verhandlungen des dritten internationalen Mathematiker-Kongresses, Heidelberg 1904 (Leipzig 1905) p. 484-491.

On the mathematical side, amazingly little is known; there have been important contributions by Oleinik, Fife, Serrin, and Chorin (references are given later), amongst many others, but the situation is far from satisfactory.

Derivation of the Boundary Layer Equations and the Boundary Layer Thickness.*

We shall begin by "deriving" the boundary layer equations for two-dimensional flow past a flat boundary. The "derivation" consists merely of plausibility arguments, which do not always hold.

Let the coordinates be x, y and the velocity field be (u, v). Let the boundary be $y = 0$ and the flow be in the upper half-plane (Figure 22-2). The Navier-Stokes equations read:

$$\left.\begin{aligned} &\frac{\partial u}{\partial t} + u\frac{\partial u}{\partial x} + v\frac{\partial u}{\partial y} = -\frac{1}{\rho}\frac{\partial p}{\partial x} + \nu\Delta u \\ &\frac{\partial v}{\partial t} + u\frac{\partial v}{\partial x} + v\frac{\partial v}{\partial y} = -\frac{1}{\rho}\frac{\partial p}{\partial y} + \nu\Delta v \\ &\operatorname{div}(u, v) = \frac{\partial u}{\partial x} + \frac{\partial v}{\partial y} = 0 \\ &\text{and} \\ &u\big|_{y=0} = 0\ , \quad v\big|_{y=0} = 0\ . \end{aligned}\right\} \qquad (1)$$

In three-dimensional space, we imagine fluid moving past a flat plate. We are interested in how the velocity field varies near the plate. Imagine that the fluid is projected toward the plate at a constant forward velocity of $(U, 0)$. It is an experimental fact that if we are not too far down the plate, the forward velocity of the fluid

*From Goldstein [2].

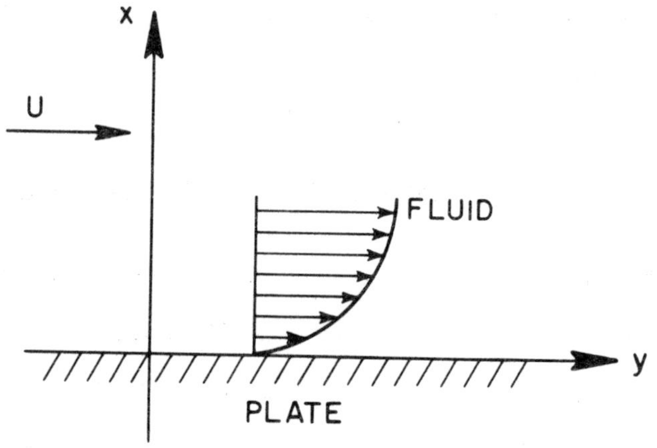

FIG. 22-2

risės rapidly from 0 at the wall to $\approx U$ in a small layer of thickness δ . The thickness of the layer grows downstream and $u(x, y) < u(x', y)$ if $x' < x$; i.e.,the flow is being retarded in its downstream motion due to friction within the fluid and the no-slip boundary condition $u|_{y=0} = 0$, $v|_{y=0} = 0$.

Consider a box on the zx-plane as shown in Figure 22-3.

Imagine that ℓ is considerably larger than δ . The difference in the forward velocities across AB and CD then should be $O(U)$; i.e.,the same order of magnitude as U . The total flow rate into/out of the box is, of course, zero. Since the flow out of CD is smaller than that into AB , there must be an upward flow across BD . The total

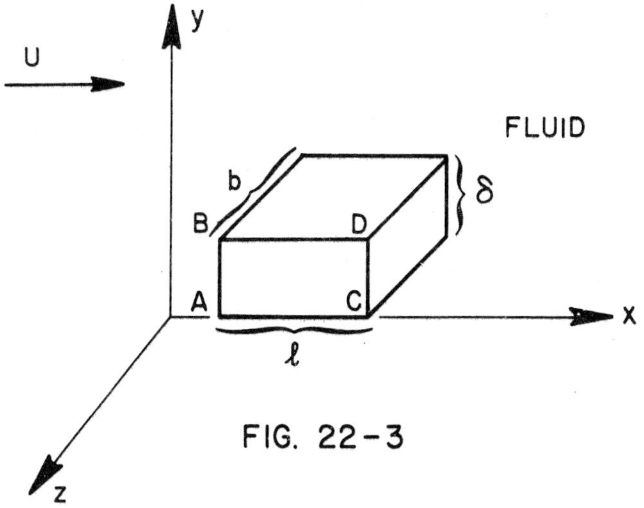

FIG. 22-3

rate of volume flow into the box across the two walls AB and CD is $O(U\ \delta b)$. Thus, the rate of flow out the wall BD must be $O(U\ \delta b)$, as there is no flow across AC . Thus

$$vb\ell = O(U\ \delta b)$$

so

$$v = O(U\ \delta/\ell).$$

Imagine choosing $\ell = O(U\ )$; then $v = O(\delta)$; i.e. the upward velocity across BD is of the same order of magnitude as δ .

We now estimate δ , the thickness of the boundary layer. The rate of forward momentum flow into/out of the box = f_x = force in the x-direction on the box. The rate of momentum flow across BD is of the order $U\,(vb\ell) = O(\rho U^2 b\delta)$. The rate of momentum flow across AB or CD is of the order $(U\,\delta b)\rho U = \rho b\delta U^2$. Thus, $f_x = O(\rho U^2 b\delta)$. Now, by assumption, the force in the boundary layer is due to the frictional forces near the wall. Recall from Section 16 that $\nu\Delta u$ = force/unit volume·density . The volume over which the force acts is $b\ell\delta$. Now $\frac{\partial^2 u}{\partial y^2} = O\,(U/\delta^2)$ and $\frac{\partial^2 u}{\partial x^2} = O\,(U/\ell^2) = O\,(\frac{1}{\ell})$. Thus $\partial^2 u/\partial x^2$ can be neglected in comparison to $\partial^2 u/\partial y^2$. Thus,

$\rho U^2 b\delta = O\,(\frac{U}{\delta}\,\nu\rho b\ell\,)$ so

$$\delta^2 = O\,(\frac{\nu\ell}{U}) \quad \text{or} \quad \delta = O\,(\frac{1}{\sqrt{Re}}) \tag{2}$$

The result (2) that the boundary layer varies as $1/\sqrt{Re}$ is an important result which is confirmed experimentally for a variety of flows.

To derive the boundary layer equations, we choose U as the standard order $(U = O(1))$. Let $\ell = O(1)$. We estimate the size of the various terms, throwing away all those not $O(1)$. The following relations hold:

$$\left.\begin{aligned} &\frac{\partial u}{\partial t} = O(1) \qquad u\,\frac{\partial u}{\partial x} = O(U^2/\ell) = O(1) \\ &v\,\frac{\partial u}{\partial y} = O(\delta U/\delta) = O(\,U\,) = O(1) \\ &\nu\,\frac{\partial^2 u}{\partial x^2} = O(\nu U/\ell^2) = O(\delta^2) \qquad \text{(throw out)} \\ &\nu\,\frac{\partial^2 u}{\partial y^2} = O(\frac{U}{\delta^2}\,\delta^2) = O(1)\ . \end{aligned}\right\} \tag{3}$$

The first equation in (1) becomes:

$$\frac{\partial u}{\partial t} + u\,\frac{\partial u}{\partial x} + v\,\frac{\partial u}{\partial y} = -\frac{1}{\rho}\,\frac{\partial p}{\partial x} + \nu\,\frac{\partial^2 u}{\partial y^2} \qquad (4)$$

Similarly we have

$$\left.\begin{aligned} \frac{\partial v}{\partial t} = O(\delta) \qquad u\,\frac{\partial v}{\partial x} &= O\left(U\,\frac{\delta}{\ell}\right) = O(\delta) \\ v\,\frac{\partial v}{\partial y} &= O\left(\delta\,\frac{\delta}{\delta}\right) = O(\delta) \\ \nu\,\frac{\partial^2 v}{\partial x^2} &= O\left(\delta^2\,\frac{\delta}{\ell^2}\right) = O(\delta^3) \\ \nu\,\frac{\partial^2 v}{\partial y^2} &= O\left(\delta^2\,\frac{\delta}{\delta^2}\right) = O(\delta)\,. \end{aligned}\right\} \qquad (5)$$

Thus, the second equation in (1) becomes merely

$$\frac{\partial p}{\partial y} = 0\,.$$

Finally $\frac{\partial u}{\partial x} = O(1) = \frac{\partial v}{\partial y}$, so there is no change in the equation $\frac{\partial u}{\partial x} + \frac{\partial v}{\partial y} = 0$.

Summarizing, we are led to the following Prandtl boundary layer equations:

$$\left.\begin{aligned} \frac{\partial u}{\partial t} + u\,\frac{\partial u}{\partial x} + v\,\frac{\partial u}{\partial y} &= -\frac{1}{\rho}\,\frac{dp}{dx}\,(x) + \nu\,\frac{\partial^2 u}{\partial y^2} \\ \frac{\partial u}{\partial x} + \frac{\partial v}{\partial y} &= 0\,. \end{aligned}\right\} \qquad (6)$$

Remarks. 1. For additional details, including the derivation for curved edges and variable U, see S. Goldstein [2].

If the boundary is curved with curvature k,

and x and y are coordinates along and normal to the wall, equations (6) are supplemented by $-ku^2 = -\frac{1}{\rho}\frac{\partial p}{\partial y}$, i.e., p now is a function of y as well as of x. For the equations in the axisymmetric case, such as the nose of an airplane, see Goldstein.

2. Equations (6) are formally deterministic for, say, $0 \leqslant x \leqslant X$, $0 \leqslant y \leqslant \delta$, $0 \leqslant t \leqslant T$ with $p(x)$ given and u, v prescribed at $t = 0$.

Existence and uniqueness theorems are complicated for these equations. (There appears to be little known for curved boundaries.) Normally one requres $u|_{x=0} > 0$, $\left.\frac{\partial u}{\partial y}\right|_{x=0} > 0$, and the results are valid only for laminar flow with Re small; the precise theorems will be stated below. (See O.A. Olenik, Mathematical Problems of Boundary Layer Theory, Soviet Math. Dokl. (1968)).

3. In order for the solutions to approximate the solutions of the Navier-Stokes equations one needs to assume, in addition that $p'(x) < 0$ (forward pressure gradient: no backflow). (See P.C. Fife, Considerations Regarding the Mathematical Basis for Prandtl's Boundary Layer Theory, Archive for Rat. Mech. and Analysis 28(1968) 184-216).

There are circumstances in which the Prandtl equations are a bad approximation and non-existence prevails.

Boundary Layer Separation.

Next, we briefly mention boundary layer separation. Consider irrotational, stationary Euler flow around a long cylinder; i.e. a two-dimensional disc (see Figure 22-4). Since $\frac{u^2}{2} + \frac{p}{\rho}$ = constant, we see that on $\widehat{AB}$, since the velocity increases in magnitude, the pressure decreases, while on $\widehat{BC}$ the velocity decreases and the pressure increases.

If a little viscosity is added, p is not changed much, but the velocity is slowed down. The backward pressure gradient then causes the forward stream to leave the surface (Figure 22-5). In practice,

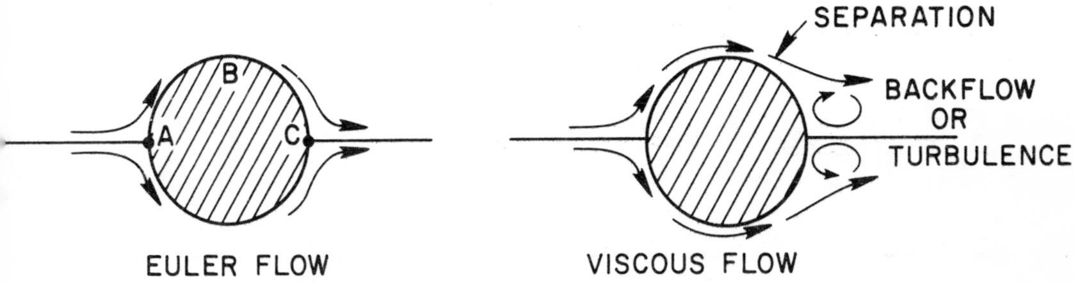

FIG. 22-4

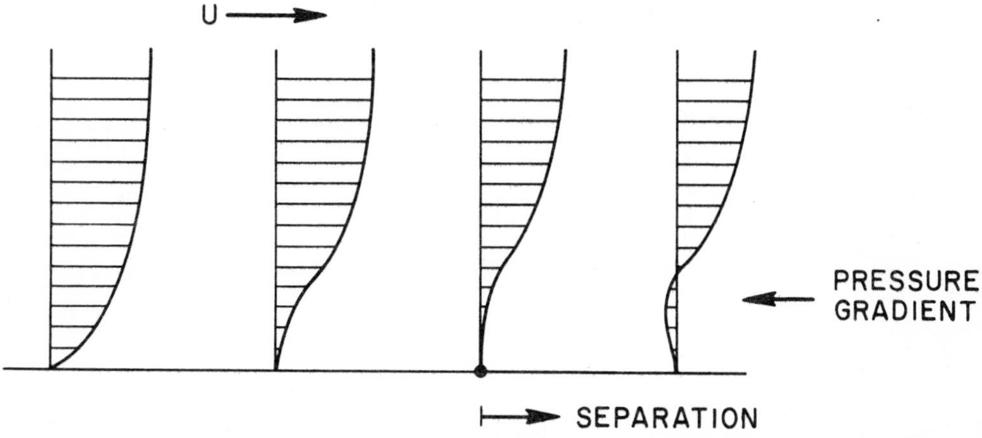

FIG. 22-5

the blow up of the Prandtl equations is often used to predict where separation occurs. (Some other methods have been proposed by A. Chorin and J. Serrin; see below.)

Existence, Uniqueness and Approximation Theorems for the Boundary Layer Equations.

We shall now go into a little more detail concerning the mathematical aspects of boundary layer theory, by stating two of the relevant theorems.

All the existence and uniqueness theorems have been proven by O.A. Oleinik. The sole approximation theorem has been proven by P.C. Fife (see the references given earlier).

Theorem (Oleinik): *Consider the equations*

$$\frac{\partial u}{\partial t} + u\frac{\partial u}{\partial x} + v\frac{\partial u}{\partial y} = -\frac{\partial p}{\partial x} + \frac{1}{R}\frac{\partial^2 u}{\partial y^2}$$

$$\frac{\partial p}{\partial y} = 0$$

$$\frac{\partial u}{\partial x} + \frac{\partial v}{\partial y} = 0$$

in the region $0 \leq y \leq \frac{1}{\sqrt{Re}}$, $0 \leq x$, $0 \leq t$ *and the boundary conditions* $u|_{y=0} = v|_{y=0} = 0$, *and* $u|_{x=0} = u_1 > 0$, u_1 *given and* $u|_{y=\frac{1}{\sqrt{Re}}} = u_2 > 0$, u_2 *given. We have initial conditions* $u|_{t=0} = u_0$ *and* $v|_{t=0} = v_0$, *given. We also have the compatibility condition*

$$-\frac{\partial p}{\partial x} = u_2 \frac{\partial u_2}{\partial x} + \frac{\partial u_2}{\partial t} .$$

If these conditions and other technical smoothness conditions are satisfied, then there is a solution (unique in a certain class of functions) to the above equations in the region $0 \leq y \leq \frac{1}{\sqrt{Re}}$, $0 \leq x \leq X$, $0 \leq t \leq T$ *for some* $X > 0$, $T > 0$. *The solution satisfies* $u > 0$.

Theorem (Fife): *Consider the equations*

$$u \frac{\partial u}{\partial x} + v \frac{\partial u}{\partial y} = -\frac{\partial p}{\partial x} + \frac{1}{R} \frac{\partial^2 u}{\partial y^2} ,$$

$$\frac{\partial p}{\partial y} = 0 ,$$

$$\text{and} \quad \frac{\partial u}{\partial x} + \frac{\partial v}{\partial y} = 0 ,$$

in the region $0 \leq y \leq \frac{1}{\sqrt{Re}}$, $0 \leq x$, $0 \leq t$. *We have boundary conditions* $u|_{x=0} = u_1 > 0$, u_1 given and $u|_{y=0} = v|_{y=0} = 0$ *and* $u|_{y=\frac{1}{\sqrt{Re}}} = u_2 > 0$. *We have the compatibility condition* $-\frac{\partial p}{\partial x} = u_2 \frac{\partial u_2}{\partial x}$.

If these conditions and other technical conditions are satisfied and if, in addition, p' *is negative, then letting* $\bar{u}$ *be the solution to the stationary Navier Stokes equations, we have*

$$|u(x, y) - \bar{u}(x, y)| \leq \frac{c}{\sqrt{Re}}$$

in the region $0 \leq x \leq X$, $0 \leq y \leq \frac{1}{\sqrt{Re}}$, *for some* $X > 0$, $c > 0$.

An Example.

Some insight into the nature of boundary layer theory can be gained from studying an example given by Serrin. (See "On the Mathematical Basis for Prandtl's Boundary Layer Theory: an Example" by James Serrin (Arch. Rat. Mech. and Analysis 28(1968) 217-225).) Consider the flow in the region shown in Figure 22-6. As mentioned in Section 16,

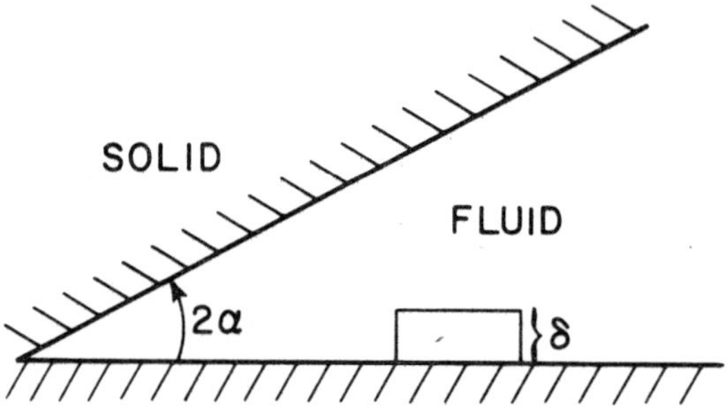

FIG. 22-6

there is a stationary solution (Hamel flow) to the Navier-Stokes equations in this region. In polar coordinates this solution takes the form

$$v_r = -\frac{f(\theta)}{r}, \quad v_\theta = 0$$

such that $f = 0$ at the walls and $f > 0$, $0 < \theta < 2\alpha$. That is, the origin is a sink into which the fluid flows at ever-increasing speed. If we write

$$v_r \cos\theta = u \ , \quad v_r \sin\theta = v$$

then (u, v) satisfies the Navier-Stokes equations if

$$\nu(f'' + 4f) - f^2 = 0 \ . \tag{7}$$

It can be shown*, without much trouble, that for sufficiently small $\nu > 0$ there are solutions f_ν with $f_\nu(\alpha) = 1$. We examine this flow as $\nu \to 0$. Formally the limiting flow ought to be $v_r = -\frac{1}{r}$, $v_\theta = 0$, an irrotational solution to the Euler equations. According to boundary layer theory, as $\nu \to 0$ in the center of the opening, $f \approx 1$ and f rises rapidly from 0 to 1 in a thin boundary layer. This is, in fact true, as can be directly verified in this case.

We look at boundary layer equations along the x-axis. The appropriate exterior speed is $U = \frac{-1}{x}$. There is a stationary solution of the boundary layer equations of the form $\bar{u} = \frac{-F(y/x)}{x}$ where

$$\nu F'' + (1 - F^2) = 0 \tag{8}$$

with boundary conditions $F(0) = 0$, $F(\infty) = 1$ and $F > 0$ (in fact $F(t) = 3\tanh^2(\frac{t}{2\nu} + c) - 2$ where $c = \operatorname{arctanh}\sqrt{2/3}$). Letting $u = v_r \cos\theta$, and $\tilde{U} = \frac{-\cos\theta}{r}$ then for $0 \leqslant \theta \leqslant \frac{3}{\sqrt{Re}}$ one can estimate**

*See S. Goldstein [2] Section 42.

**See Serrin's paper for details. The proof rests on the comparison method for ordinary differential equations.

that

$$\left|\frac{u-\bar{u}}{\tilde{U}}\right| \leq \frac{32}{Re} ,$$

i.e., we have a good approximation to the boundary layer equations in the boundary layer. In the central opening:

$$\sqrt{\tfrac{1}{2}Re}\ \log Re \leq \theta \leq 2\alpha - \sqrt{\tfrac{1}{2}Re}\ \log Re \quad \text{and} \quad V \equiv \frac{-\sin\theta}{r}$$

$$\left|\frac{v-V}{V}\right| \leq \frac{24}{Re} ,$$

i.e., we approximate Euler flow outside the boundary layer.

The cases of a source and sink are quite different. A sink at high Reynolds number is stable whereas a source is unstable (see Section 18). This example is, however, "simple" in the sense that boundary layer separation and turbulence do not occur.

More on Boundary Layer Separation.

We shall discuss some conditions for boundary layer separation.* Suppose we are dealing with stationary flows and that $\lim_{\nu\to 0} \nu/\delta^2 = 0$. In deriving the Prandtl equations we would then get just

---

*See J. Serrin "Mathematical Aspects of Boundary Layer Theory" (Lecture Notes, Univ. of Minnesota 1962) for additional details, and a discussion of the case $\lim_{\nu\to 0} \nu/\delta^2 > 0$.

$$u \frac{\partial u}{\partial x} + v \frac{\partial u}{\partial y} = -\frac{1}{\rho} \frac{dp}{dx} ,$$

$$\text{and} \quad \frac{\partial u}{\partial x} + \frac{\partial v}{\partial y} = 0 .$$

But at the wall $(y = 0)$, u and v vanish, so $p'(x) = 0$.

Then $\frac{\partial(u/v)}{\partial y} = \frac{\frac{\partial u}{\partial y} v - \frac{\partial v}{\partial y} u}{v^2} = \frac{-u\frac{\partial u}{\partial x} - u\frac{\partial v}{\partial y}}{v^2} = 0$. So $v = f(x)u$, and the equations are merely $\frac{\partial u}{\partial x} + f(x) \frac{\partial u}{\partial y} = 0$. Streamlines are easily checked to be either the points where $u = v = 0$, horizontal lines or curves $y = \int_0^x f(s)ds + C$. Thus we get two possibilities for streamlines, as shown in Figure 22-7.

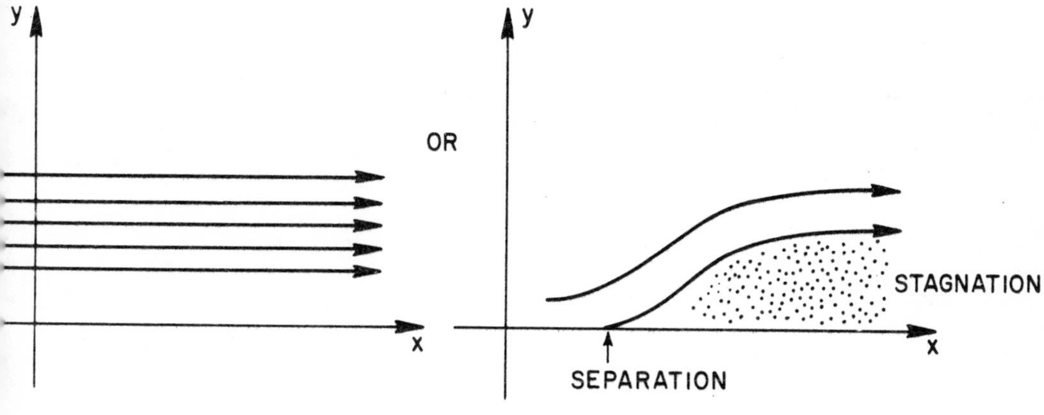

FIG. 22-7

Boundary layer theory for compressible flow has been developed quite far. For example, see the important article Multistructured Boundary Layers on Flat Plates and Related Bodies by K. Stewartson, Advances in Applied Mechanics, Vol. 14, Academic Press (1974).

General features of separation were discussed earlier. However, useful conditions which can be used in examples to predict separation are not so simple. Chorin has recently proposed a numerical method for solving the Navier-Stokes equations, which will be briefly discussed in the next section. There is promise that the development of separation (see Figure 22-5) can be numerically detected using this method.

Separation and turbulence complicate the theory and intuitive understanding of lift and drag. Again it is possible that numerical computation (using sophisticated mathematical models) is the only recourse. The drag coefficient (related to drag D and cross-sectional area A normal to the flow by $D = C_D(\frac{\rho V^2}{2})A$) for a cylinder is shown in Figure 22-8.

Research Problem. Try to make sense out of this section.

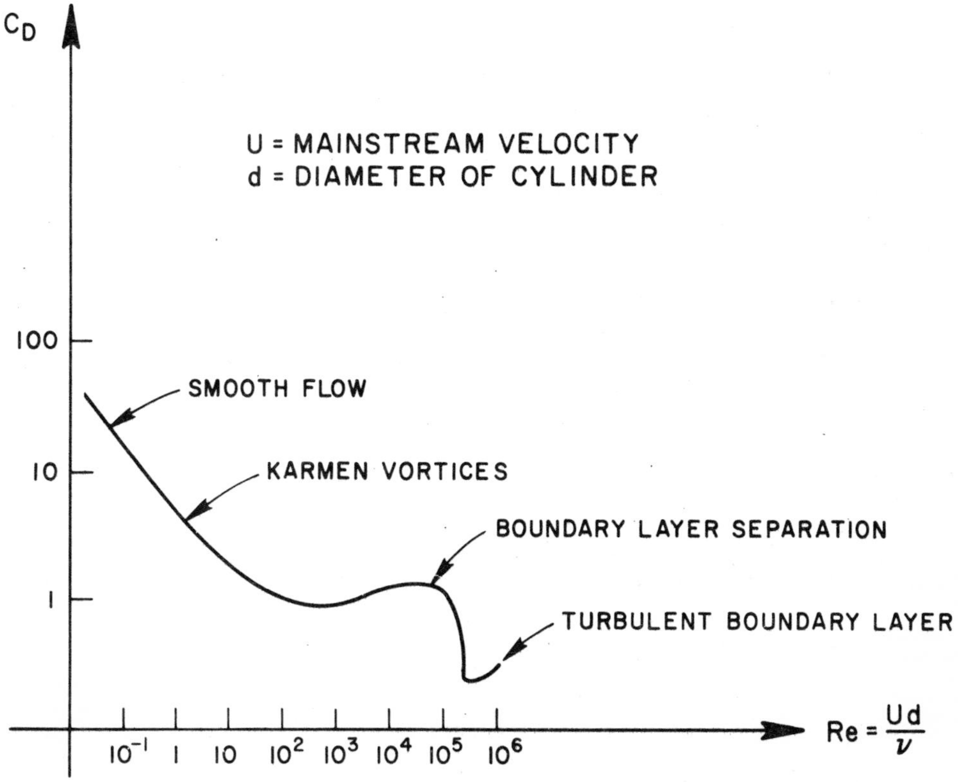

C_D
U = MAINSTREAM VELOCITY
d = DIAMETER OF CYLINDER
100
10
1
SMOOTH FLOW
KARMEN VORTICES
BOUNDARY LAYER SEPARATION
TURBULENT BOUNDARY LAYER
10^{-1} 1 10 10^2 10^3 10^4 10^5 10^6
$Re = \frac{Ud}{\nu}$

FIG. 22-8

§23. Remarks on Turbulence Theory.

It has to be admitted that of all the areas of fluid mechanics, turbulence is the least understood. There are great masses of literature on the subject, but most are concerned with computational methods or statistical results under various ad hoc hypotheses, which often turn out to be oversimplified. There are very few results which go into the nature of turbulence, its qualitative features and how the statistical properties arise.

Here we shall merely mention some of the basics of the subject and give a few speculations. For further information, see the following references:

[5] L.D. Landau, L.M. Lifschitz, "Fluid Mechanics," Addison Wesley (1959),

[17] P.G. Saffman, Lectures on Homogeneous Turbulence in "Topics in Nonlinear Physics", N. Zabuski, Ed. (Springer 1968),

[14] A. Chorin, "Lectures on Turbulence Theory", Publish or Perish (1975).

Turbulence is usually defined operationally as fluid motion that looks chaotic or random. It usually is thought to really be so, and this kind of blanket randomness assumption is at the source of many of the difficulties. Indeed, recent experimental evidence indicates much more fine structure to turbulence than was heretofore believed. (See the following articles in the Journal of Fluid Mechanics: G. Brown and A. Roshko 64(1974) 775, C.D. Winant and F.K. Browand 63(1974) 237, C.T. Lau and M.J. Fisher, 67(1975) 299, P.G. Mulhearn and R.E. Luxton 68 (1975) 577 and H. Ueda and J.O. Minze 67(1975) 125.)

The theory of turbulence is in an extremely primitive state. For example no one has shown theoretically that solutions of the Navier-Stokes equations exist which correspond to turbulent flows we see; nor has any one shown that even the simplest model for turbulence--homogeneous turbulence, in which mean properties are independent of position--can exist in any precise sense.

Turbulence occurs in flow around bodies, both near the boundary ("boundary layer turbulence") and downstream ("free turbulence") as well as in more innocent looking flows such as patches of turbulence in the atmosphere.

Turbulence apparently has four key features

(a) some sort of randomness,

(b) diffusive nature (such as vortex lines stretching and twisting),

(c) dissipation of kinetic energy into heat,

and (d) the dominance of the nonlinear terms in the flow ("strong nonlinearity").

Kolmogorov's Law.

The velocity correlation tensor of homogeneous random flow is defined by

$$R_{ij}(\underline{r}, t) = \overline{u_i(t, \underline{x})u_j(t, \underline{x} + \underline{r})}$$

where $\overline{\qquad\qquad}$ indicates an average of some unspecified sort is often introduced to study turbulence. The variable t is usually suppressed.

Its Fourier transform,

$$\phi_{ij}(t, \underline{k}) = \frac{1}{(2\pi)^3} \int R_{ij}(t, \underline{r}) e^{-i\underline{k}\cdot\underline{r}} \, d\underline{r}$$

is the <u>energy spectrum tensor</u>.

Notice that the mean kenetic energy is given by

$$\tfrac{1}{2}\, \overline{v^2} = \int \tfrac{1}{2}\, \phi_{ii}(\underline{k}) d\underline{k} \qquad \text{(sum on } i \text{)}$$

so that $\phi_{ii}(\underline{k})$ gives the density of the contribution to kinetic energy in wave number space $\underline{k}$. The <u>energy spectrum function</u> is defined by

$$E(k) = \int_{|\underline{k}|=k} \tfrac{1}{2}\, \phi_{ii}(\underline{k}) dA$$

so that

$$\tfrac{1}{2}\, \overline{v^2} = \int_0^\infty E(k) dk \ .$$

A key intuitive ingredient is the following: energy flows (or cascades) from large eddies (small k) to small eddies (large k) where it dissipates in the form of heat due to the viscosity.

Energy is thought of as being pumped into the fluid through large eddy motion at some constant rate ε , and dissipated by viscosity ν in the small eddies; see Figure 23-1.

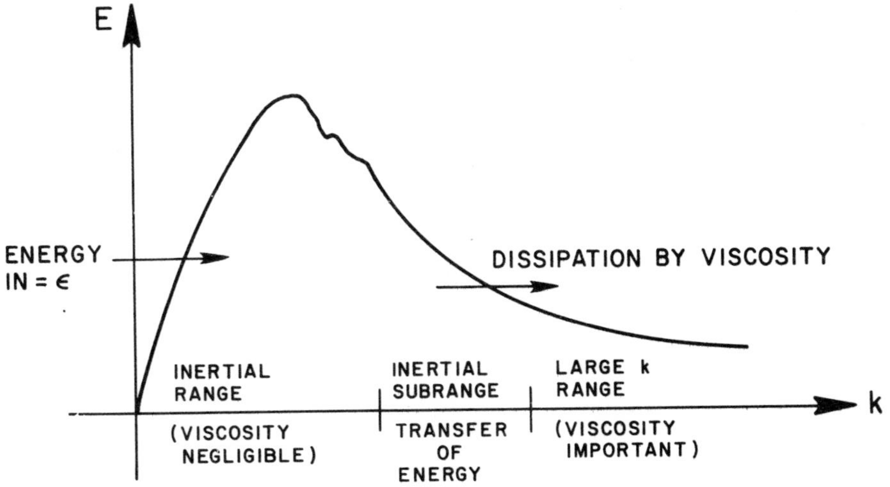

FIG. 23-1

Kolmogorov* assumed that for large k , $E(k)$ should depend only on ε and k ; furthermore, he assumed the small scale motion is isotropic and its statistical properties depend only on ε and ν .

From ε and ν one forms

$$\ell = \left(\frac{\nu^3}{\varepsilon}\right)^{1/4} \quad \text{a length}$$

and

$$V = (\nu\varepsilon)^{1/4} \quad \text{a velocity}$$

by dimensional considerations. Note $V = (\varepsilon\ell)^{1/3}$; so one says that velocities of eddies of size ℓ are proportional to $(\varepsilon\ell)^{1/3}$.

Statistical quantities are assumed to depend just on ℓ and V , in a dimensionally correct fashion. Thus we should have

---

*A.N. Kolmogorov, C.R. Acad. Sci. 30(1941) 301 and 32(1941) 16.

$$E(k) = CV^2\ell(k\ell)^\alpha$$

for constants C and α (k has the dimensions of length^{-1}) .

In the "inertial subrange" where energy transfer occurs, $E(k)$ is assumed to depend only on ε and k. The combination in which V^2 and fit to give a quantity independent of ν is

$$V^2\ell^{-2/3} = \varepsilon^{2/3}$$

Thus we should choose $\alpha = -5/3$ and get

$$E(k) = C\varepsilon^{2/3}\, k^{-5/3}$$

which is the famous Kolmogorov law. It is more or less experimentally ve fied in three dimensions but is inappropriate in two dimensions. For a c cal and detailed examination of this law, we recommend the works of Saffman and Chorin mentioned above.

Kolmogorov's law is intriguing because it gives insight into the nature of turbulence and may give clues to the nature of solutions of the Navier-Stokes equations. In particular, it may give insight into whether or not solutions remain regular for all time, or blow up in a finite time; this is called the problem of global regularity.*

---

*These remarks are based on work of Chorin, Frisch and others. See J. Marsden, D. Ebin, A. Fischer, Diffeomorphism Groups, Hydrodynamics and Relativity, Proc. 13th Bienniel Seminar of Can. Math. Congress 1972 p. 207 and articles in the proceedings of the Orsay conference on turbulence, June 1975 (Springer Lecture Notes).

The Nature of Turbulence.

The flavor of the statistical approach was indicated above. However, it does not face squarely the problem of how turbulence arises, how it is related to exact solutions of the Navier-Stokes equations and how the statistical nature of turbulence can arise.

Now we shall briefly sketch a bold program of Ruelle and Takens† to tackle this. The key starting point is the Hopf bifurcation theorem mentioned in Section 18. Recall that this result showed how a stable periodic point can be replaced by an attracting periodic orbit. Further bifurcations may replace this orbit by an attracting 2-torus, and so on. Ruelle and Takens have argued that in this or other situations, complicated ("strange") attractors can be expected and that this lies at the roots of the explanation of turbulence.*

In the particular case where tori of increasing dimension form, the model is a technical improvement over an earlier idea of E. Hopf wherein turbulence results from a loss of stability through successive branching. It seems however that strange attractors may form in other cases too, such as in the Lorenz equations which are used in studies of atmospheric flow.** This is perfectly consistent with the general Ruelle-

---

†On the Nature of Turbulence, Comm. Math. Phys. 20 (1971) 167-192, 23 (1971) 343-4.

*Some people have also tried to explain the statistics in terms of ergodic properties using the Hamiltonian nature of Euler's equations. This seems to be quite a complex program (see Chorin [14] for some discussion.)

**See Marsden-McCracken "The Hopf Bifurcation" Springer Notes in Appl. Math. (1976), §4B.

Takens picture, as are the closely related "snap through" ideas of Joseph and Sattinger (Bifurcating Time Periodic Solutions and Their Stability, Arch. Rat. Mech. An. 45 (1972) 79-109).

In the branching process, stable solutions become unstable as the Reynolds number is increased. Hence turbulence is supposed to be a necessary consequence of the equations and, in fact, of the "generic case" and just represents a complicated solution. For example in the Couette flow (see Section 12) as one increases the angular velocity Ω_1 of the inner cylinder, one finds a shift from laminar flow to Taylor cells or related patterns at some bifurcation value of Ω_1. Eventually turbulence sets in. In this scheme, as has been realized for a long time, one first looks for a stability theorem and for when stability fails. For example, if one stayed close enough to laminar flow, one would expect the flow to remain approximately laminar. Serrin has a theorem of this sort which we present as an illustration:

Stability Theorem. *Let* $\Omega \subset \mathbb{R}^3$ *be a bounded domain and suppose the flow* $\underline{v}_t^\nu$ *is prescribed on* $\partial\Omega$ *(this corresponds to having a moving boundary, as in Couette flow). Let* $V = \max_{\underline{x}\in\Omega,\, t\geq 0} \|\underline{v}_t^\nu(\underline{x})\|$, d = *diameter of* Ω *and* ν *equal the viscosity. Then if the Reynolds number* $Re = (Vd/\nu) \leq 5.71$, $\underline{v}_t^\nu$ *is universally* L^2 *stable among solutions of the Navier-Stokes equations.*

Universally L^2 stable means that if $\tilde{\underline{v}}_t^\nu$ is any other solution to the equations with the same boundary conditions, then the L^2

norm (or energy) of $\tilde{\underline{v}}_t^\nu - \underline{v}_t^\nu$ goes to zero as $t \to 0$.

The proof is really very simple and we recommend reading Serrin [11] for the argument.

Chandresekar, Serrin, Velte and many other authors have analyzed criteria of this sort in some detail for Couette flow.

As a special case, we recover something that we expect. Namely if $\underline{v}_t^\nu = 0$ on $\partial\Omega$ is any solution for $\nu > 0$ then $\underline{v}_t^\nu \to 0$ as $t \to \infty$ in L^2 norm, since the zero solution is universally stable.

A traditional definition (as in Landau-Lifschitz [5]) says that turbulence develops when the vector field $\underline{v}_t$ can be described as $\underline{v}_t(w_1, \ldots, w_n) = \underline{f}(tw_1, \ldots, tw_n)$ where $\underline{f}$ is a quasi-periodic function, i.e., $\underline{f}$ is periodic in each coordinate, but the periods are not rationally related. For example, if the orbits of the $\underline{v}_t$ on the tori given by the Hopf theorem can be described by spirals with irrationally related angles, then $\underline{v}_t$ would be such a flow.

Considering the above example a bit further, it should be clear there are many orbits that the $\underline{v}_t$ could follow which are qualitatively like the quasi-periodic ones but which fail themselves to be quasi-periodic. In fact a small neighborhood of a quasi-periodic function may fail to contain many other such functions. One might desire the functions describing turbulence to contain most functions and not only a sparse subset. More precisely, say a subset U of a topological space S is <u>generic</u> if it is a Baire set (i.e., the countable intersection of open dense subsets). It seems reasonable to expect that the functions describing turbulence should be generic, since turbulence is a common phenomenon and the equations of flow are never exact. Thus we would want a theory of turbulence that would not be destroyed by adding on small perturbations to the equations of motion.

The above sort of reasoning lead Ruelle-Takens to point out that since quasi-periodic functions are not generic, it is unlikely they "really" describe turbulence. In its place, they propose the use of "strange attractors." These exhibit much of the qualitative behavior one would expect from "turbulent" solutions to the Navier-Stokes equations and they are stable under perturbations of the equation; i.e., are "structurally stable".

For an example of a strange attractor, see S. Smale Differentiable Dynamical Systems, Bull. Am. Math. Soc. 73(1967) 747-817. Usually strange attractors look like (Cantor sets) × (manifolds) , at least locally.

Ruelle and Takens have shown that if we define a strange attractor A to be an attractor which is neither a closed orbit nor a point, and disregarding non-generic possibilities such as a "figure 8," then there ar strange attractors on T^4 in the sense that a whole open neighborhood of vector fields has a strange attractor as well.

If the attracting set of the flow, in the space of vector fields which is generated by the Navier-Stokes equations is strange, then a solution attracted to this set will clearly behave in a complicated, turbulent manner. While the whole set is stable, individual points in it are not. Thus (see Figure 23-2) an attracted orbit is constantly near unstable (nearly periodic) solutions and gets shifted about the attractor in an aimless manner. Thus we have the following reasonable definition of turbulence as proposed by Ruelle-Takens:

" ... the motion of a fluid system is turbulent when this motion is described by an integral curve of a vector field $\underline{X}_\mu$ which tends to a set A , and A is neither empty nor a fixed point nor a closed orbit."

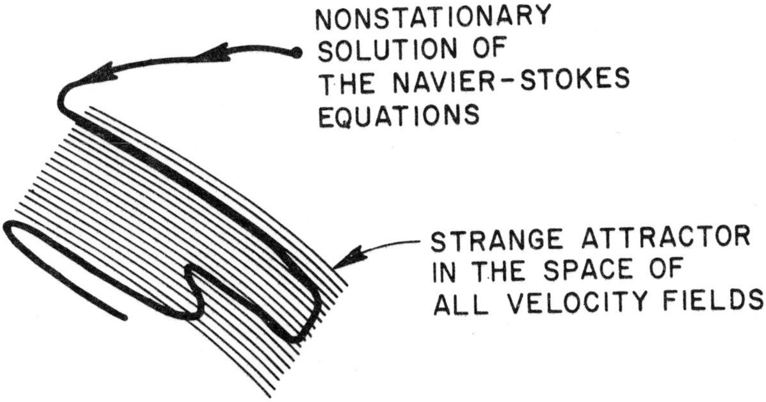

FIG. 23-2

Recently Ruelle and Bowen* have shown how the usual statistical mechanics of ergodic systems can be used to study the case of strange attractors, following work of Bowen and Sinai. It remains to connect this up with observed statistical properties of fluids, like the time average of the pressure in turbulent flow.

* See R. Bowen, " Equilibrium States and the Ergodic Theory of Anosov Diffeomorphisms", Springer Lecture Notes #470 (1975).

In summary then, this view of turbulence may be phrased as follows. Our solutions for small μ (= Reynolds number in many fluid problems) are stable and as μ increases, these solutions become <u>unstable</u> at certain critical values of μ and the solution falls to a more complicated <u>stable</u> solution; eventually, after a certain (finite) number of such bifurcations, the solution falls to a strange attractor (in the space of all time dependent solutions to the problem). Such a solution, which is wandering close to a strange attractor, is called <u>turbulent</u>.

The fall to a strange attractor may occur after a Hopf bifurcation to an oscillatory solution and then to invariant tori, or may appear by some other mechanism, such as in the Lorenz equations as explained above ("snap through turbulence").

Global Regularity.

Leray, in his classic papers in the 1930's*, argued that the Navier-Stokes equations might break down and the solutions fail to be smooth when turbulence ensues. This idea was amplified when Hopf in 1950 proved global existence (in time) of <u>weak</u> solutions to the equations, but not uniqueness. It was speculated that turbulence occurs when strong changes to weak and uniqueness is lost.† However, it is still unknown whether or not this really can happen (Hopf's theorem is proved in Ladyzhenskaya [12].)

*Acta Mathematica <u>63</u>(1934) 193-248.

†The set of times where the solution is not regular was studied in detail by Leray. This set, which might be called "turbulence times," is apparently a closed set of Hausdorff dimension $\leqslant \frac{1}{2}$ (V. Schaeffer), although of course it may be empty.

The Ruelle-Takens and Leray pictures are in conflict. Indeed, if strange attractors are the explanation, their attractiveness implies that solutions remain smooth for all t. Indeed, from the work on the Hopf bifurcation, one can show that near the stable closed orbit solutions are defined and remain smooth and unique for all $t \geqslant 0$. This is already in the range of interesting Reynolds numbers where global smoothness is not implied by the classical estimates.

It is known that in two dimensions the solutions of the Euler and Navier-Stokes equations are global in t and remain smooth. In three dimensions it is unknown and is called the "global regularity" or "all time" problem.

Recent numerical evidence* seems to suggest that the answer is negative for the Euler equations. It is anybody's guess for the Navier-Stokes equations.

We wish to make two points in the way of conjectures. 1. In the Ruelle-Takens picture, global regularity for all initial data is not an *a priori* necessity; the basins of the attractors will determine which solutions are regular and will guarantee regularity for turbulent solutions (which is what most people now believe is the case).
2. Global regularity, if true in general, will probably never be proved by making estimates on the equations. One needs to examine in much more depth the attracting sets in the infinite dimensional dynamical system of the Navier-Stokes equations and to obtain the *a priori* estimates this way.

---

*Orsay conference on turbulence, June 1975, Springer Lecture Notes (to appear).

Two major open problems:

(i) Identify a strange attractor in a specific flow of the Navier-Stokes equation (e.g. pipe flow, flow behind a cylinder, etc.).

(ii) Link up the ergodic theory on the strange attractor with the statistical theory of turbulence.

Chorin's Formula

The problem of obtaining turbulent solutions theoretically as well as numerically and information on the limit $Re \to \infty$ (or $\nu \to 0$) for flow past obstacles may be aided by a recent method due to Chorin.* We will briefly explain this formula now.

If $\underline{A}$ and $\underline{B}$ are $n \times n$ matrices, then

$$e^{t(\underline{A}+\underline{B})} = \lim_{n\to\infty}\left(e^{t\underline{A}/n} e^{t\underline{B}/n}\right)^n \tag{1}$$

The generalization of this formula to general Lie groups or to semi-groups of operators is usually called the Lie-Trotter product formula.

If $\underline{X}$ and $\underline{Y}$ are smooth vector fields on $\mathbb{R}^n$ (or a manifold) with flows $\underline{F}_t$ and $\underline{G}_t$ respectively, then the flow $\underline{H}_t$ of $\underline{X} + \underline{Y}$, (maximally defined), is given by

$$\underline{H}_t(\underline{x}) = \lim_{n\to\infty}(\underline{F}_{t/n} \circ \underline{G}_{t/n})^n(\underline{x}) \tag{2}$$

i.e., to solve $\frac{d\underline{x}}{dt} = \underline{X}(\underline{x}) + \underline{Y}(\underline{x})$, we may use the solutions of $\frac{d\underline{x}}{dt} = \underline{X}(\underline{x})$ and $\frac{d\underline{x}}{dt} = \underline{Y}(\underline{x})$ separately and employ the time step iteration

*Cf. J. Marsden, A Formula for the Solution of the Navier-Stokes Equations Based on a Method of Chorin, Bull. Am. Math. Soc. 80(1974) 154-158, A. Chorin, Numerical Study of Slightly Viscous Flow, J. Fluid Mech. 57(1973) 785-796.

procedure (2). This reduces to (1) if $\underline{X}(\underline{x}) = \underline{A}\cdot\underline{x}$, $\underline{Y}(\underline{x}) = \underline{B}\cdot\underline{x}$. The method (2) is implicit and well known in numerical analysis for evolution equations.

It is tempting to apply (2) to the Navier-Stokes equations where $\underline{F}_t$ is the flow of the Stokes equation and $\underline{G}_t$ the flow of the Euler equations. For manifolds with no boundary this method works quite well and was used by Ebin and Marsden in 1970 to show the convergence of the solutions of the Navier-Stokes equations to the Euler equations as the viscosity $\nu \to 0$ (convergence in H^s , any large s). The motivation for using (2) is that the Stokes equation and Euler equation enter in a symmetrical way and we do not use the Laplace operator to dominate the nonlinear terms, a process that can lead to complications for small ν .

In the case of a boundary, Chorin has proposed a numerical scheme which seems to work quite well numerically in 2 or 3 spatial dimensions and has errors which are bounded as $\nu \to 0$. (Although some may differ on the execution of the scheme, it is the over-all method which we are stressing as important here). Translated into a product formula, this scheme reads as follows.

$$H_t(\underline{v}) = \lim_{n\to\infty}(F_{t/n} \circ \Phi_{t/n} \circ G_{t/n})^n \underline{v} \tag{3}$$

Here: $\underline{v}$ is a divergence free vector field, $\underline{v} = \underline{0}$ on the boundary $\partial\Omega$ of the region Ω in question

: G_t is the flow of the Euler equations (boundary conditions $\underline{v}$

parallel to $\partial\Omega$)

: F_t is the semiflow of the Stokes equation (boundary conditions $\underline{v} = \underline{0}$ on $\partial\Omega$)

: Φ_t is the "vorticity creation operator" which maps a $\underline{v} \parallel \partial\Omega$ to a $\underline{v} = \underline{0}$ on $\partial\Omega$ by adding on a vorticity field to $\underline{v}$ whose backflow cancels $\underline{v}$ on $\partial\Omega$ (the vorticity field is constructed as in Ladyzhenskaya [12, p. 24-5])

: H_t is the semiflow of the Navier-Stokes equations.

[The full details of (3) are not yet proven. The reason is that we find it convenient to use the Sobolev function space $W^{2,p}(\Omega)$ with zero boundary conditions ($p > 3$ in three dimensions). In such a space the Stokes equations are quite difficult and the needed L_p estimates are only now becoming understood.]

However, we expect to be able to use (3) to prove:

for fixed initial data $\underline{v}_0$, there exists a $T > 0$ independent of ν (limited only by the time of existence for the Euler equations), a time of existence for regular solutions of the Navier-Stokes equations and these solutions converge in L_p to those of the Euler equations as $\nu \to 0$, $0 \leqslant t < T$.

Formula (3) will converge for as large a t as $H_t(\underline{v})$ is known to exist and be regular. Thus, if $H_t(\underline{v})$ represents a turbulent, but regular solution of the Navier-Stokes equation, (3) will remain valid. Indeed, Chorin has obtained some good numerical results for very high Reynolds numbers.

What this formula (3) does is to make explicit the old intuitive idea that vorticity is created on the boundary because of the difference in the boundary conditions between the Euler and Navier-Stokes equations and that if the Reynolds number is high this vorticity is swept downstream by the Euler flow to form the turbulent wake.

It would be most beautiful to make some link between this picture and the Ruelle-Takens picture of turbulence mentioned above.

The reader can find many references on turbulence which are worthwhile reading besides those already mentionsed. Two particularly interesting ones are Recent Theories of Turbulence by J. Von Neumann (Collected Works VI, Macmillan (1963) 437-472) and "Les Objects Fractals" by B. Mandelbrot, (Flammarion, Paris (1975)).

The depth, difficulty, and controversy about turbulence--its mechanism and structure--makes it one of the most challenging areas of all of fluid mechanics.

Appendices

A. Some Vector Identities.

1. $$\begin{cases} \underline{\nabla}(f+g) = \underline{\nabla}f + \underline{\nabla}g \\ \underline{\nabla}\cdot(\underline{v}+\underline{w}) = \underline{\nabla}\cdot\underline{v} + \underline{\nabla}\cdot\underline{w} \\ \underline{\nabla}\times(\underline{v}+\underline{w}) = \underline{\nabla}\times\underline{v} + \underline{\nabla}\times\underline{w} \end{cases}$$

2. $$\begin{cases} \underline{\nabla}\cdot(f\underline{v}) = \underline{\nabla}f\cdot\underline{v} + f\underline{\nabla}\cdot\underline{v} \\ \underline{\nabla}\times(f\underline{v}) = \underline{\nabla}f\times\underline{v} + f\underline{\nabla}\times\underline{v} \end{cases}$$

3. $$\begin{cases} \underline{\nabla}\cdot(\underline{v}\times\underline{w}) = \underline{w}\cdot\underline{\nabla}\times\underline{v} - \underline{v}\cdot\underline{\nabla}\times\underline{w} \\ \underline{\nabla}\times(\underline{v}\times\underline{w}) = (\underline{w}\cdot\underline{\nabla})\underline{v} - \underline{w}(\underline{\nabla}\cdot\underline{v}) - (\underline{v}\cdot\underline{\nabla})\underline{w} + \underline{v}(\underline{\nabla}\cdot\underline{w}) \end{cases}$$

4. $$\underline{\nabla}(\underline{v}\cdot\underline{w}) = (\underline{w}\cdot\underline{\nabla})\underline{v} + (\underline{v}\cdot\underline{\nabla})\underline{w} + \underline{w}\times(\underline{\nabla}\times\underline{v}) + \underline{v}\times(\underline{\nabla}\times\underline{w})$$

5. $$\begin{cases} \underline{\nabla}\times\underline{\nabla}f = \underline{0} \\ \underline{\nabla}\cdot(\underline{\nabla}\times\underline{v}) = 0 \end{cases}$$

6. $$\Delta\,\underline{v} = \underline{\nabla}(\underline{\nabla}\cdot\underline{v}) - \underline{\nabla}\times(\underline{\nabla}\times\underline{v})$$

7. $$(\underline{v}\cdot\underline{\nabla})\underline{v} = \frac{1}{2}\underline{\nabla}v^2 - \underline{v}\times(\underline{\nabla}\times\underline{v})$$

B. Vector Formulae and the Navier-Stokes Equations in Cylindrical and Spherical Coordinates.

Cylindrical.

With notation as in Figure A-1, the following formulae hold:

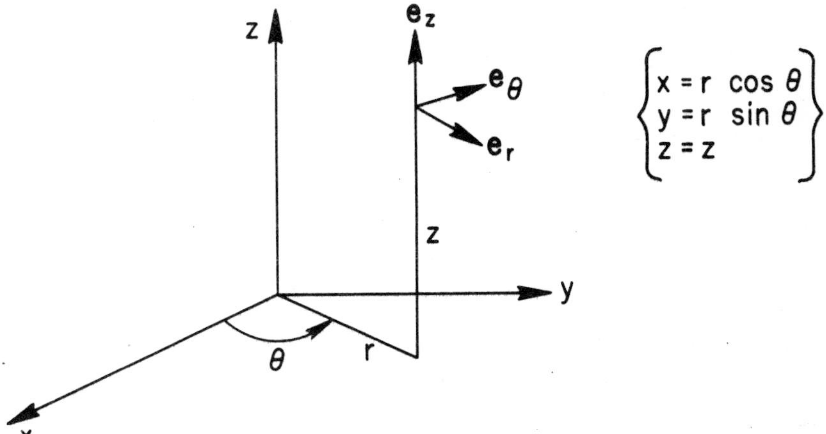

FIG. A-1

(C1) $\underline{\nabla} f = \frac{\partial f}{\partial r}\underline{e}_r + \frac{1}{r}\frac{\partial f}{\partial \theta}\underline{e}_\theta + \frac{\partial f}{\partial z}\underline{e}_z$

(C2) $\underline{\nabla}\cdot\underline{v} = \frac{1}{r}\{\frac{\partial}{\partial r}(rv_r) + \frac{\partial}{\partial \theta}v_\theta + \frac{\partial}{\partial z}(rv_z)\}$

(C3) $\underline{\nabla}\times\underline{v} = \frac{1}{r}\begin{vmatrix} \underline{e}_r & r\underline{e}_\theta & \underline{e}_z \\ \frac{\partial}{\partial r} & \frac{\partial}{\partial \theta} & \frac{\partial}{\partial z} \\ v_r & rv_\theta & v_z \end{vmatrix}$

(C4) $\Delta f = \frac{1}{r}\frac{\partial}{\partial r}(r\frac{\partial f}{\partial r}) + \frac{1}{r^2}\frac{\partial^2 f}{\partial \theta^2} + \frac{\partial^2 f}{\partial z^2}$

(C5) $\frac{Df}{Dt} = \frac{\partial f}{\partial t} + v_r\frac{\partial f}{\partial r} + \frac{v_\theta}{r}\frac{\partial f}{\partial \theta} + v_z\frac{\partial f}{\partial z}$

(C6) Navier-Stokes Equations (incompressible):

$$\rho \frac{D\underline{v}}{Dt} = \rho \{ \frac{\partial \underline{v}}{\partial t} + (\underline{v} \cdot \underline{\nabla})\underline{v} \} = -\underline{\nabla} p + \rho \underline{f} + \mu \Delta \underline{v}$$

become:

(a) $$\rho [\frac{Dv_r}{Dt} - \frac{v_\theta^2}{r}] = \rho f_r - \frac{\partial p}{\partial r} + \mu [\Delta v_r - \frac{v_r}{r^2} - \frac{2}{r^2} \frac{\partial v_\theta}{\partial \theta}]$$

(b) $$\rho [\frac{Dv_\theta}{Dt} + \frac{v_r v_\theta}{r}] = \rho f_\theta - \frac{1}{r} \frac{\partial p}{\partial \theta} + \mu [\Delta v_\theta + \frac{2}{r^2} \frac{\partial v_r}{\partial \theta} - \frac{v_\theta}{r^2}]$$

(c) $$\rho \frac{Dv_z}{Dt} = \rho f_z - \frac{\partial p}{\partial z} + \mu \Delta v_z$$

with $\frac{D}{Dt}$ and Δ as in (C5) and (C4).

Spherical.

With notation as in Figure A-2, the following formulae hold:

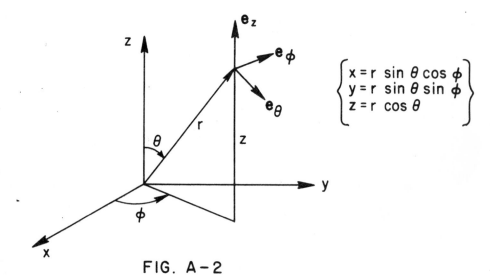

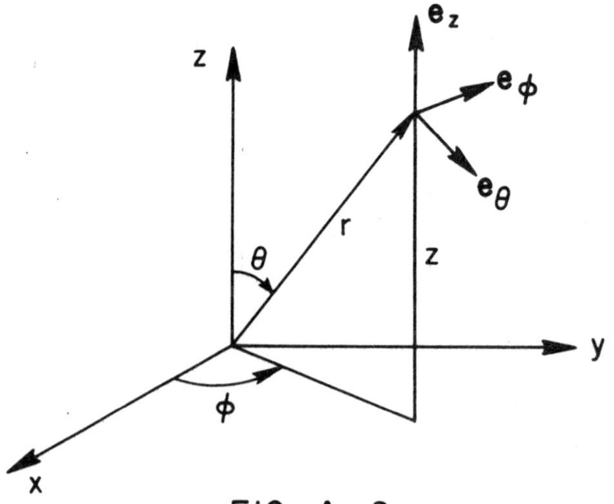

$$\begin{cases} x = r \sin\theta \cos\phi \\ y = r \sin\theta \sin\phi \\ z = r \cos\theta \end{cases}$$

FIG. A-2

(S1) $$\underline{\nabla} f = \frac{\partial f}{\partial r} \underline{e}_r + \frac{1}{r} \frac{\partial f}{\partial \theta} \underline{e}_\theta + \frac{1}{r \sin\theta} \frac{\partial f}{\partial \phi} \underline{e}_\phi$$

(S2) $$\underline{\nabla}\cdot\underline{v} = \frac{1}{r^2}\frac{\partial}{\partial r}(r^2 v_r) + \frac{1}{r\sin\theta}\frac{\partial}{\partial\theta}(v_\theta \sin\theta) + \frac{1}{r\sin\theta}\frac{\partial v_\phi}{\partial\phi}$$

(S3) $$\underline{\nabla}\times\underline{v} = \frac{1}{r^2\sin\theta}\begin{vmatrix} \underline{e}_r & r\underline{e}_\theta & r\sin\theta\,\underline{e}_\phi \\ \frac{\partial}{\partial r} & \frac{\partial}{\partial\theta} & \frac{\partial}{\partial\phi} \\ v_r & rv_\theta & r\sin\theta\, v_\phi \end{vmatrix}$$

(S4) $$\Delta f = \frac{1}{r^2}\frac{\partial}{\partial r}\left(r^2\frac{\partial f}{\partial r}\right) + \frac{1}{r^2\sin\theta}\frac{\partial}{\partial\theta}\left(\sin\theta\frac{\partial f}{\partial\theta}\right) + \frac{1}{r^2\sin^2\theta}\frac{\partial^2 f}{\partial\phi^2}$$

(S5) $$\frac{Df}{Dt} = \frac{\partial f}{\partial t} + v_r\frac{\partial f}{\partial r} + \frac{v_\theta}{r}\frac{\partial f}{\partial\theta} + \frac{v_\phi}{r\sin\theta}\frac{\partial f}{\partial\phi}$$

(S6) Navier-Stokes equations (incompressible)

$$\rho\frac{D\underline{v}}{Dt} = \rho\left\{\frac{\partial\underline{v}}{\partial t} + (\underline{v}\cdot\underline{\nabla})\underline{v}\right\} = -\underline{\nabla}p + \rho\underline{f} + \mu\Delta\underline{v}$$

become:

(a) $$\rho\left[\frac{Dv_r}{Dt} - \frac{v_\theta^2+v_\phi^2}{r}\right] = \rho f_r - \frac{\partial p}{\partial r} + \mu\left[\Delta v_r - \frac{2v_r}{r^2} - \frac{2}{r^2}\frac{\partial v_\theta}{\partial\theta} - \frac{2v_\theta\cot\theta}{r^2} - \frac{2}{r^2\sin\theta}\frac{\partial v_\phi}{\partial\phi}\right]$$

(b) $$\rho\left[\frac{Dv_\phi}{Dt} + \frac{v_\phi v_r}{r} - \frac{v_\phi^2\cot\theta}{r}\right] = \rho f_\theta - \frac{1}{r}\frac{\partial p}{\partial\theta} + \mu\left[\Delta v_\theta + \frac{2}{r^2}\frac{\partial v_r}{\partial\theta} - \frac{v_\theta}{r^2\sin^2\theta} - \frac{2\cos\theta}{r^2\sin^2\theta}\frac{\partial v_\phi}{\partial\phi}\right]$$

(c) $$\rho\left[\frac{Dv_\phi}{Dt} + \frac{v_\phi v_r}{r} + \frac{v_\theta v_\phi\cot\theta}{r}\right] = \rho f_\phi - \frac{1}{r\sin\theta}\frac{\partial p}{\partial\phi} + \mu\left[\Delta v_\phi - \frac{v_\phi}{r^2\sin^2\theta} + \frac{2}{r^2\sin^2\theta}\frac{\partial v_r}{\partial\phi} + \frac{2\cos\theta}{r^2\sin^2\theta}\frac{\partial v_\theta}{\partial\phi}\right]$$

with $\frac{D}{Dt}$ and Δ as in (S5) and (S4).

C. The Stress Tensor for a Viscous Incompressible Fluid.

Cartesian: $T_{ij} = -p\delta_{ij} + \mu\left(\frac{\partial v_i}{\partial x_j} + \frac{\partial v_j}{\partial x^i}\right)$

Cylindrical:

$$T_{rr} = -p + 2\mu\left(\frac{\partial v_r}{\partial r}\right) \qquad T_{r\theta} = \mu\left(\frac{1}{r}\frac{\partial v_r}{\partial \theta} + \frac{\partial v_\theta}{\partial r} - \frac{v_\theta}{r}\right) \qquad T_{rz} = \mu\left(\frac{\partial v_z}{\partial r} + \frac{\partial v_r}{\partial z}\right)$$

$$T_{\theta\theta} = -p + 2\mu\left(\frac{1}{r}\frac{\partial v_\theta}{\partial \theta} + \frac{v_r}{r}\right) \qquad T_{\theta z} = \mu\left(\frac{\partial v_\theta}{\partial z} + \frac{1}{r}\frac{\partial v_z}{\partial \theta}\right)$$

$$T_{zz} = -p + 2\mu\left(\frac{\partial v_z}{\partial z}\right)$$

Spherical:

$$T_{rr} = -p + 2\mu\left(\frac{\partial v_r}{\partial r}\right) \qquad T_{r\phi} = \mu\left(\frac{\partial v_\phi}{\partial r} + \frac{1}{r \sin\theta}\frac{\partial v_r}{\partial \phi} - \frac{v_\phi}{r}\right) \qquad T_{r\theta} = \mu\left(\frac{1}{r}\frac{\partial v_r}{\partial \theta} + \frac{\partial v_\theta}{\partial r} - \frac{v_\theta}{r}\right)$$

$$T_{\phi\phi} = -p + 2\mu\left(\frac{1}{r \sin\theta}\frac{\partial v_\phi}{\partial \phi} + \frac{v_r}{r} + \frac{v_\theta \cot\theta}{r}\right) \qquad T_{\theta\phi} = \mu\left(\frac{1}{r \sin\theta}\frac{\partial v_\theta}{\partial \phi} + \frac{1}{r}\frac{\partial v_\phi}{\partial \theta} - \frac{v_\phi \cot\theta}{r}\right)$$

$$T_{\theta\theta} = -p + 2\mu\left(\frac{1}{r}\frac{\partial v_\theta}{\partial \theta} + \frac{v_r}{r}\right)$$